THE BODY IN BRIEF
Essentials for Healthcare

Rebecca Rayman, R.N., B.S.N.

Illustrated

Skidmore-Roth Publishing, Inc.
El Paso, Texas

SR
Skidmore-Roth

Cover Design: Robert Pawlak

Library of Congress Cataloging-in-Publication Data

Rayman, Rebecca.
The body in brief. 2nd ed.

Includes bibliographies and index.
1. Nursing -- Outlines, syllabi, etc. I. Title.
[DNLM: 1. Anatomy -- nurses' instruction. 2. Nursing Assessment.
3. Physiology -- nurses' instruction.

ISBN 0-944132-76-6

THE BODY IN BRIEF
Table of Contents

HOW TO USE THIS BOOK

The Body in Brief, Essentials for Health Care, was designed as a quick reference to provide information on topics which are needed in the everyday clinical practice. The book was also designed with portability in mind, so that it would fit into the pocket of a lab coat, backpack or purse. Topics for inclusion were chosen carefully concentrating on information that would be useful to the student as well as the practitioner.

The book is divided into nine chapters based on major systems of the body. Each chapter is further divided into subsections. The subsections are:

I. Overview
II. Assessment
III. Laboratory and diagnostic tests
IV. Procedures and conditions
V. Diets
VI. Drugs
VII. Glossary

How to Use Each Section

Section I -- OVERVIEW

The overview contains information on the primary functions of the organ system as a whole unit as well as the components of each system. Drawings are included. This section will also include the physiology of the system. The purpose of this section is to provide a review of the system. Where appropriate flow sheets are included to provide a visual progression of the material discussed.

Section II -- ASSESSMENT

This section includes two subsections, the health history and the physical assessment. This section is designed to aid the clinician, whether beginning or with experience to perform a thorough assessment of the system. The health history and physical examination forms are provided as an assessment tool. The health history assessment forms include open-ended questions to assess the patient's health beliefs. A space to write notes is included on the physical assessment form.

Section III -- LABORATORY AND DIAGNOSTIC TESTS

The section on laboratory and diagnostic tests provides information on the more common tests used in the system. The tables include information on expected results, why the test may have been ordered (indications), collection information and what abnormal results may signify. The diagnostic test information tables provide the expected results, indications, and what should be done before and after the procedure.

Section IV -- PROCEDURES AND CONDITIONS

Although labeled "Procedures and Conditions" this section contains information pertaining to the system that would fall under a variety of headings. Information here was chosen based on what would be helpful to the student or nurse in the clinical setting. Drawings and flow sheets are included where needed to supplement the text.

Section V -- DIETS

Information on diets that may be ordered for the patient with problems involving this system. Each diet includes why it would be ordered, what foods are restricted, what foods are allowed and interventions for that diet. This section should be helpful when the patient wants a late night snack and the dietitian is not available, or when the patient or family asks about the diet.

Section VI -- DRUGS

This section contains information on drug classifications used in this system and gives examples of commonly-used medications.

Section VII -- GLOSSARY

Included to help decipher unfamiliar words within the text or to look up words that are unfamiliar when reading medical reports or charts.

I would also like to express my appreciation to Lillian Mayberry, Ph.d., who carefully reviewed the second edition.

Rebecca J. Rayman, R.N., B.S.N.

Cardiovascular System

CARDIOVASCULAR SYSTEM

Table of contents

Section I -- OVERVIEW

Primary functions

- Transport oxygen, nutrients and other substances to the cells
- Transport and remove carbon dioxide, produced by cellular metabolism via the lungs and cellular wastes via the kidneys
- Aid in the regulation of body temperature

Components and functions

The primary components of the cardiovascular system are the heart and blood vessels of the body. Figure 1A shows the heart and Figure 1D shows the primary blood vessels

Heart

The heart is the primary organ of the circulatory system. The heart is a pump that forces the blood throughout the body. The heart has two sides; the right side and the left side are separated by the interventricular septum

Vena cavae

The vena cavae are veins that lead into the heart from the body. All blood returning to the heart enters through one of these two veins. Note in Figure 1A there are no valves where these veins empty into the right atrium

Superior vena cava

The superior vena cava returns unoxygenated blood from the head and upper portion of the body to the heart

Inferior vena cava

The inferior vena cava is where unoxygenated blood returns from the lower portion of the body to the heart

Chambers

The heart has four chambers; each side of the heart has two chambers. The chambers which receive the blood are known as the atria. The chambers that force blood out of the heart are the ventricles

Atria

There are no valves at the entry points to the atrial chambers so blood flows in continuously. The atria have the capacity to hold 57 ml or two ounces of blood

Right atrium

Unoxygenated blood flows into the right atrium from the superior and inferior vena cava. The right atrium is larger and its wall is thinner than the left atrium

STRUCTURE OF THE HEART

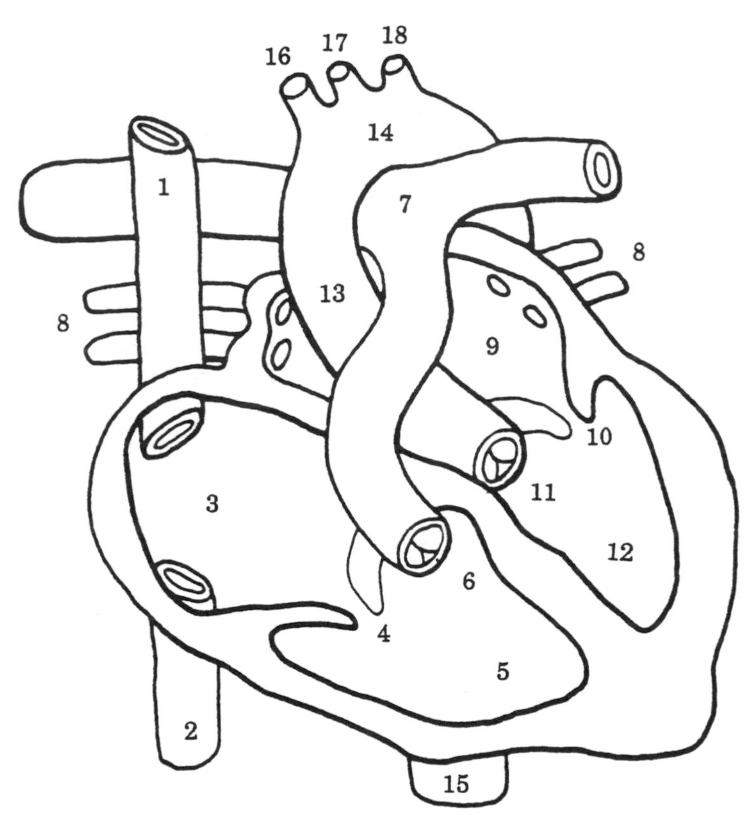

Figure 1A

1. Superior vena cava
2. Inferior vena cava
3. Right atrium
4. Tricuspid valve
5. Right ventricle
6. Pulmonary valve
7. Pulmonary artery
8. Pulmonary veins
9. Left atrium
10. Bicuspid (mitral) valve
11. Aortic valve
12. Left ventricle
13. Ascending aorta
14. Aortic arch
15. Decending thoracic aorta
16. Brachiocephalic artery
17. Left common carotid artery
18. Left subclavian artery

Left atrium

Oxygenated blood flows into the left atrium from the pulmonary veins on its way back to the heart from the lungs.

Ventricles

The ventricles receive blood from the atria via the atrioventricular (AV) valves and then force the blood through the semilunar valves (pulmonary and aortic) into arteries. Because the ventricles force blood out, their walls must be able to generate and withstand more pressure than the walls of the atria which receive blood passively. Therefore, ventricle walls are thicker. The ventricles hold 85 ml or about three ounces.

Right ventricle

The right ventricle has the tricuspid valve at its entrance and the pulmonary valve (semilunar valve) at its exit. The right ventricle has unoxygenated blood flowing through it on the way to the lungs. Like the right atrium, the right ventricle wall is only about one-third as thick as the left ventricle wall.

Left ventricle

The left ventricle has the bicuspid valve at its entrance and the aortic valve at its exit. This blood has been oxygenated in the lungs and the left ventricle will pump it to the body.

Heart valves

Blood travels through the heart in one direction because of the valves. There are one-way valves at the exit point of each chamber to prevent blood from flowing back. The valves open and close in response to pressure changes.

AV valves

The valves between the chambers on either side of the heart are known as the atrioventricular, or AV, valves. They are the tricuspid and bicuspid.

Tricuspid valve

The AV valve on the right side has three cusps or flaps and is known as the tricuspid valve. When these triangular cusps are open blood flows into the right ventricle.

Bicuspid valve

The AV valve on the left side has two cusps and is known as the bicuspid valve. This is also called the mitral valve.

Semilunar valves

The semilunar valves are located in the arteries at the exit point of the ventricles. The semilunar valves prevent the blood from flowing back to the heart.

Pulmonary valve
The semilunar valve located between the pulmonary artery and the right ventricle is the pulmonary valve.
Aortic valve
The semilunar valve located between the aorta and the left ventricle is the aortic valve.

Blood flow and the heart

- The heart is a 4-chambered, 2-sided pump; each side pumps simultaneously. On both sides of the heart, the atria contract just prior to contraction of the ventricles
- The right side of the heart forces oxygen-deficient blood through the lungs where it picks up oxygen
- The left side of the heart forces oxygen-rich blood (returning from the lungs) to the body where the oxygen will be used by the cells
- Blood flows in one direction because the heart valves open and close with contraction and relaxation of the chambers

  Systole (Greek-contraction) is the period of time when the ventricles contract

  Diastole (Greek-to expand) is the relaxation phase of the ventricle when the chambers are refilling with blood
- Stroke volume (SV) is the amount of blood ejected by a ventricle during systole (averages 70 cc)
- The average heart rate (HR) is approximately 72 times per minute
- Cardiac output (CO) equals the amount of blood ejected from the left ventricle into the aorta per minute
- CO = SV times beats per minute. Therefore:

  CO = SV X HR
  CO = 70 cc X 72 beats per minute
  CO = 5,040 cc/min, or approximately 5 liters
- The AV valves are open during diastole and closed during systole. It is the closure of the AV valves that makes the first heart sound (S1), heard as lubb
- The semilunar valves are open during systole and closed during diastole. Closure of the semilunar valves makes the second heart sound (S2), heard as dubb
- The heart requires a constant supply of oxygenated blood and receives its own supply from the coronary arteries which branch from the aorta and encircle the myocardium

CIRCULATION OF BLOOD THROUGH THE HEART

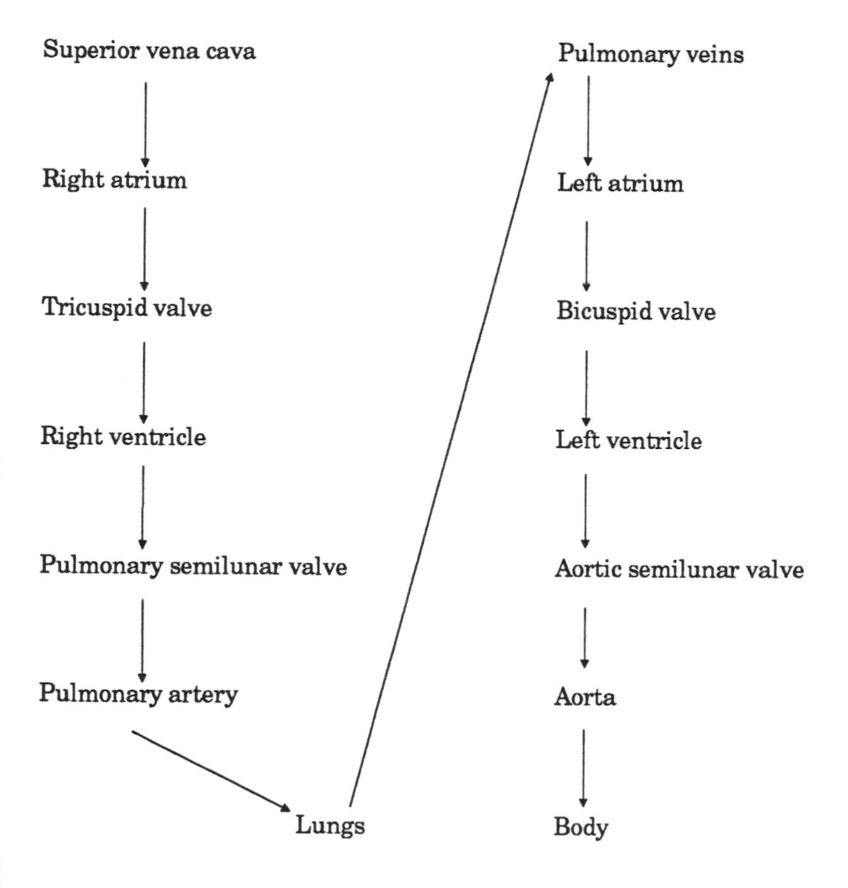

Figure 1B

The heart's electrical conduction system

The heart has a system known as the electrical conduction system to initiate the cardiac cycle. (Figure 1C).

The cardiac cycle is the period from the beginning of one heart beat to the beginning of the next heart beat. The components of the heart's electrical conduction system are:

Sinoatrial node (SA node)
Atrioventricular node (AV node)
Bundle of His
Purkinje fibers

- The SA node is the pacemaker of the heart. The cardiac cycle begins when this node fires (depolarizes)
- The electrical impulse (depolarization wave) generated by the SA node travels through the atrial tissue to the AV node
- This same impulse stimulates the atria to contract
- When the atria contract, blood is forced into the ventricles
- Once the impulse reaches the AV node there is a short delay. It is then rapidly transmitted to the AV bundle or Bundle of His
- The Bundle of His divides into the left and right bundle branches which pass down the interventricular septum
- From the bundle branches the impulse is transmitted to the Purkinje fibers which penetrate into deeper myocardium
- When the impulse reaches the ventricular muscle, contraction occurs and the blood is forced out of the heart to the lungs or body
- The cardiac cycle is complete, and the SA node is ready to fire again
- If the SA node does not function, the AV node takes over the initiation of the cardiac cycle but at a slower rate

Blood flow within the body

- The amount of blood needed by the tissues and organs is variable depending on the body's demand for oxygen
- A common example of the body's increased need for oxygen is exercise
- The amount of blood flowing through the body can be regulated to meet the body's needs
- Blood flow can be increased by increasing the heart rate and/or stroke volume. This would increase the volume and rate at which blood is circulated
- Blood flow can be increased by regulating the amount of blood any one organ receives, such as increasing the blood flow to the

Electrical Conduction of the Heart

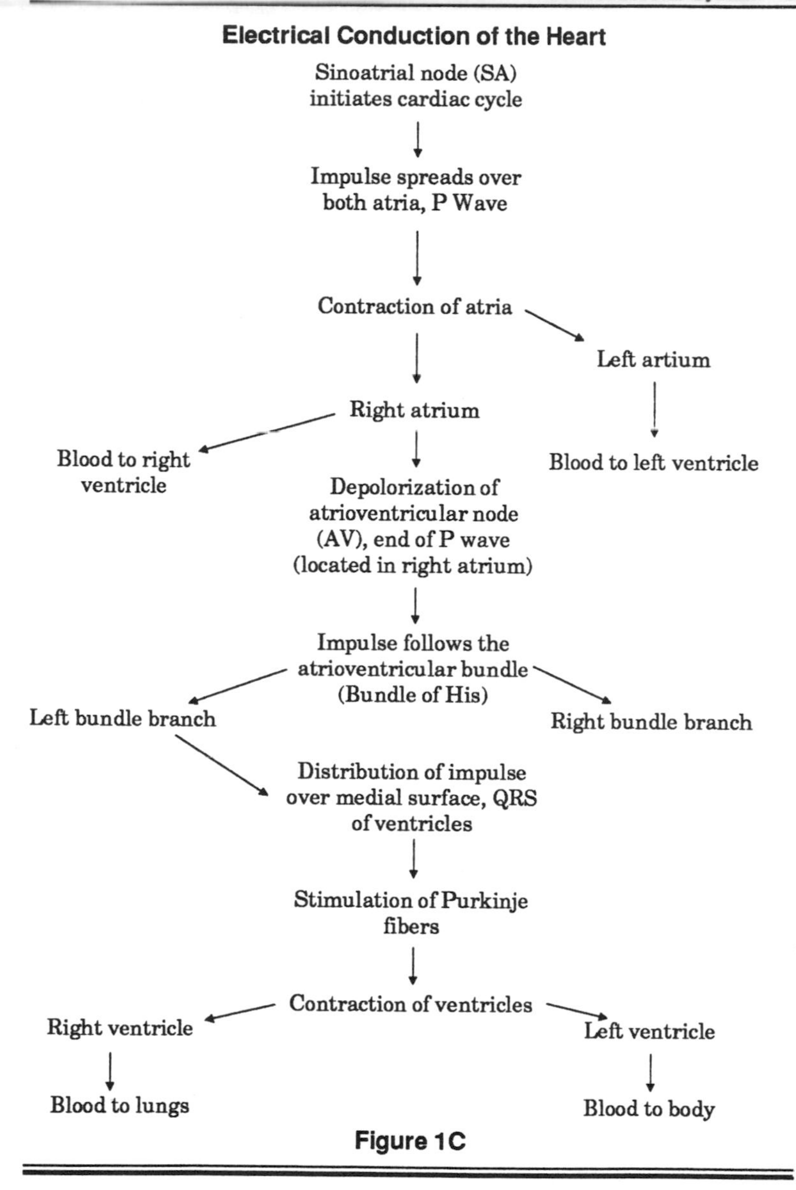

Figure 1C

vital organs and simultaneously decreasing blood flow to non-essential organs.

The blood vessels

There are two types of blood vessels: arteries and veins. Arteries carry blood from the heart to the tissues. Larger arteries lead into smaller arterioles and then finally to capillaries where the exchange of most nutrients and wastes occur. All arteries except pulmonary arteries carry oxygenated blood. Veins return blood to the heart. Blood flow back to the heart begins in the capillaries, then passes into the venules and finally into the larger veins. All veins except pulmonary veins carry unoxygenated blood. See Figure 1D.

The major arteries of the body

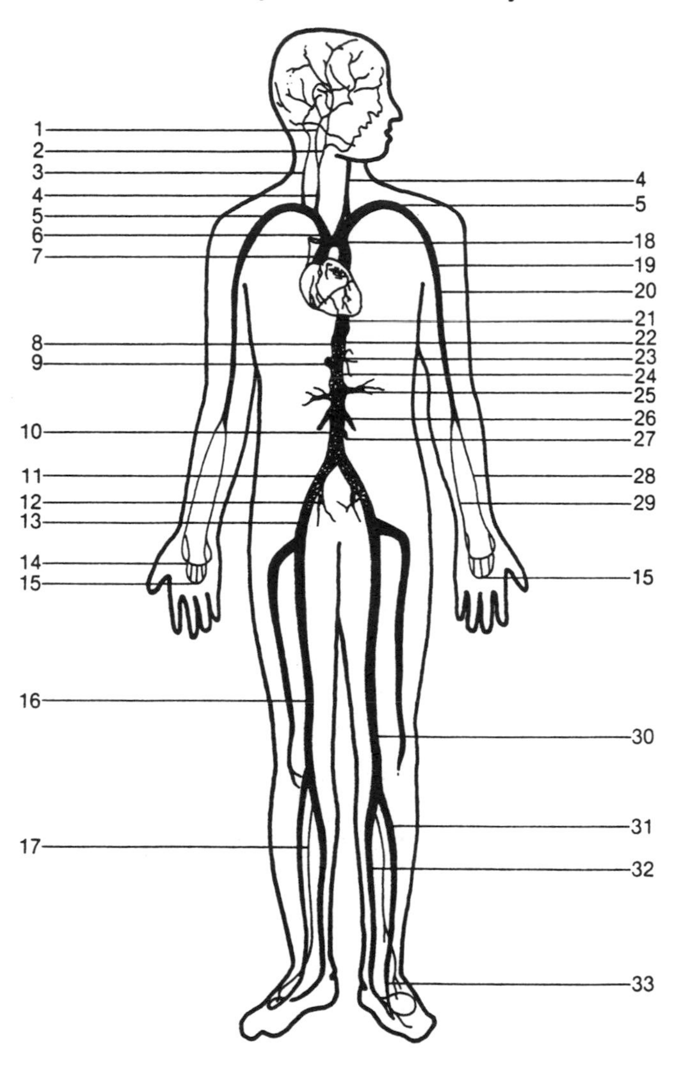

Figure 1D

1. Internal carotid -- supplies blood to the brain, eyes, and nose.

2. External carotid -- supplies blood to the neck, face, and skull.

3. Vertebral -- supplies blood to the muscles of the neck, vertebrae, spinal cord, and brain.

4. Common carotid -- supplies blood to the neck and thyroid.

5. Subclavian arteries -- supply blood to the brain, meninges, spinal cord, neck, and arms.

6. Brachiocephalic -- supplies blood to right common carotid and the subclavian arteries.

7. Ascending aorta -- supplies blood to the heart.

8. Celiac -- supplies blood to the liver, stomach, pancreas, and spleen.

9. Common hepatic -- supplies blood to the liver.

10. Abdominal aorta -- supplies blood to the celiac artery and the legs.

11. Common iliacs -- supply blood to the pelvis, abdominal wall, and the legs.

12. Internal iliacs -- supply blood to the uterus, bladder, and muscles to the buttocks and thighs.

13. External iliacs -- supply blood to the abdominal wall, external genitalia (sex organs), and lower extremities.

14. Deep palmar arch -- supplies blood to the palm and fingers.

15. Superficial palmar arch -- supplies blood to the palm and fingers.

16. Femoral -- supplies blood to the external genitalia, abdomen, and legs.

17. Peroneal -- supplies blood to the muscles of the calf and ankle.

18. Arch of the aorta -- supplies blood to the brachiocephalic, carotids, and subclavin arteries.

19. Axillary -- supplies blood to the underarm and the brachial artery.

20. Brachial -- supplies blood to the upper arms.

21. Thoracic aorta -- supplies blood to the chest muscles and esophagus.

22. Gastric -- supplies blood to the stomach.

23. Splenic -- supplies blood to the spleen, pancreas, and stomach.

24. Superior mesenteric -- supplies blood to the intestines and colon.

25. Renal - supplies blood to the kidneys and ureters.

27. Inferior mesenteric -- supplies blood to the colon and rectum.

26. Testicular/Ovarian -- supplies blood to the colon and rectum

28. Radial -- supplies blood to the forearm and palm arteries.

29. Ulnar -- supplies blood to the forearm, wrist, and hand.

30. Popliteal -- supplies blood to the knees and calf.

31. Anterior tibial -- supplies blood to the leg, ankle, and foot.

32. Posterior tibial -- supplies blood to the leg, foot, and heel.

33. Dorsalis pedis -- supplies blood to the ankle and foot.

Section II -- ASSESSMENT

Taking a health history and performing a physical assessment are the keys to a cardiovascular assessment.

Health History

Chief complaint

Examples of some common complaints include:

Chest pain
Shortness of breath
Rapid or slow pulse
Heart palpitations
Dizziness, fainting
High blood pressure
Intolerance to exercise

Personal and family history

Because family history is a significant factor in heart problems, it is important to ask: Is there a family history of heart disease?

High blood pressure/rheumatic fever/heart attack or surgery

For rheumatic fever -- Assess if patient is aware of any heart valve damage.

Heart attack (myocardial infarction, MI) -- List date, time in hospital, required lifestyle changes.

Heart surgery -- List date, name or type, surgeon, hospital and required lifestyle changes.

Varicose veins -- List corrective surgery.

Chest pain

Chest pain is a common chief complaint. It does not always indicate a heart attack, although chest pain is the most significant symptom. Pain may radiate to shoulder, jaw, and down the left arm. If chest pain is present assess location, severity, and duration. Ask the patient to describe the pain and radiation patterns.

Pain that is deep, dull, and not localized may be angina, myocardial infarction, tumor, aortic aneurysm, pulmonary embolism, or gallbladder disease.

Pain that comes and goes is more likely to be angina. Pain that is severe and persistent is more likely to be a heart attack.

Pain that is sharp and localized may be pleurisy, pneumothorax, pericarditis, hiatal hernia, breast lesions, rib fracture, costochondritis, or gallbladder disease.

Pain that intensifies by deep breathing or coughing and can be localized may be pleurisy, broken ribs, or a torn muscle.

Heart palpitations

Palpitations are a rapid, throbbing or fluttering of the heart. Palpitations are common and have a variety of causes.

Shortness of breath/swelling in ankles or feet

These signs indicate the heart's failure to pump adequately.

Cardiovascular system testing

Assess if patient has any cardiovascular testing, and if so, when? Why was testing done? Assess patient's knowledge and understanding of outcome of testing.

CARDIOVASCULAR ASSESSMENT

Chief complaint

Patient's statement________________Onset________________________________

Frequency____________Duration_________Other areas affected__________________

Have you had this before?______________Date of last episode____________________

What treatment was given?____

What do you think caused this?____

What changes have been made in lifestyle because of this problem?__________________

____

Personal and family history

	Patient	Family member
High B/P	_______	_____
Rheumatic fever	_______	_____
Heart attack	_______	_____
Heart surgery	_______	_____
Varicose veins	_______	_____
Leg ulcers	_______	_____
Blood clots	_______	_____

	Patient Only	
	Current	Past
Palpitations	________	________
Shortness of breath	________	________
Swelling in feet/ankles	________	________
Arm/leg pain	________	________
Numbness in extremity	________	________
Syncope	________	________
Chest pain	________	________
Where	________	________
Severity	________	________
Frequency	________	________
Duration	________	________
Describe	________	________

Cardiovascular system testing

Electrocardiogram______Echocardiogram________Coronary arteriogram____________

Angiogram______Cardiac cath________Stress test______Holter monitor____________

Venogram_______Ultrasound_______Chest X-ray________Blood tests_____________

Current treatments/medications

Who is your physician?___________________________Phone____________________

Medication: ____________________________Dose______Frequency_____Route______

Use of oxygen__________How much____________Frequency______________________

Cigarettes/day______/______yrs. Alcohol/kind__________Amount_____/_____yrs.

Usual weight___________Type of exercise______________________Frequency________

What is your usual pulse?__________________Blood pressure_________/__________

How far can you walk comfortably?____

Physical Assessment

Physical assessment of the cardiovascular system is done in the following order:

1. Inspection
2. Palpation
3. Auscultation

Assessment of the heart, abdominal aorta, peripheral pulse and blood pressure are important. Patient is undressed to the waist for this exam. Patient should be assessed in supine, left lateral, and sitting positions.

Inspection

General appearance

What position has the patient assumed? If the patient is short of breath he may best tolerate a sitting position. If the patient is having chest pain he may have their hand on their chest or even be splinting the chest with a pillow. How does the patient look? Someone experiencing acute chest pain or heart failure will appear anxious and restless. The patient may complain of weakness or feeling faint.

General skin color/mucous membranes/nailbeds

Is the patient pale, sweating, flushed or cyanotic?

Pale, cool, moist skin with frank sweating (diaphoresis) are cardinal signs of a heart attack.

Cyanosis can occur in heart failure or other disorders where there is a lack of oxygen. Check mucous membranes and nailbeds.

Flushed skin can occur with any condition that leads to vasodilation.

Apical impulse

When patient is supine, is the apical impulse visible? Stand to the right side and observe the chest. Can you see any pulsations, heaves or lifts? If so, indicate location using the intercostal spaces as landmarks. Usually the apical impulse is located at the midclavicular line in the fifth intercostal space. The apical impulse is visible in most adults. If the chest heaves or lifts with the heart beat, it indicates that the heart is working harder than normal.

Clubbing of fingers

Clubbing of fingers is indicative of inadequate oxygenation and is often related to chronic heart or lung disease.

Varicose veins

Often seen on the veins of the lower extremities -- indicates stretching of the vascular wall due to increased venous pressure. This results when the venous valves that prevent back flow or pooling of venous blood are not functioning properly.

Palpation

Use the fingertips or the ball of your hand during palpation. Avoid startling the patient with cold hands, rub them together to warm them. When palpating, begin at the midclavicular line in the fifth intercostal space on the left side. The patient should be supine.

Cardiac thrills

A cardiac thrill is a series of vibrations often related to a loud cardiac murmur. It has been described as the purring of a cat. Record the precise location. A thrill is always indicative of pathology.

Radial/carotid/femoral/popliteal/dorsalis pedis

Check each pulse point bilaterally, are they equal? What is the rate? Record: Left radial pulse=80 (RL 80)

Auscultation

Listen at the four valve locations first with the stethoscope diaphragm, then with the bell. The diaphragm picks up high pitched sounds best like S1, and S2. The bell is for low-pitched sounds S3 and S4. Apply the bell lightly, the diaphragm firmly. It may be necessary to auscultate the chest with the patient in more than one position to determine the presence of murmurs.

Apical pulse

The apical pulse is at the fifth intercostal space near the midclavicular line. Listen for one full minute. The first sound is S1 (lubb). This signals the closure of the bicuspid and tricuspid valves. The second sound S2 (dubb) is the closure of the aortic and pulmonic valves.

Is the rate regular? If not, is there a pattern? Does it vary with respirations?

What is the intensity: faint, moderate, loud?

Apical/radial pulse

An apical/radial pulse is taken by two people simultaneously. One records the apical pulse, the other the radial pulse for one full minute.

Four heart valves

Listen to see which sound is louder in each of the four valular areas. Record which heart sound is louder, S1 or S2. Are there extra heart sounds?

S3/S4

The third heart sound, S3, is best heard at the apex of the heart with the bell of the stethoscope. S3 follows S2 and is normal in children and young adults, but is usually abnormal in older adults. The fourth heart sound, S4, is heard best at the apex of the heart and occurs shortly before S1; it is best heard with the bell. S4 may indicate coronary artery disease and/or hypertension.

Murmurs

A murmur occurs when blood flow through the heart is abnormal. A murmur is often heard when there is a malfunctioning valve which may allow blood to flow backwards into a heart chamber. Listen for murmurs, a murmur will last longer than the heart sounds normally heard. Grade the murmur:

1=Very faint, difficult to hear
2=Quiet
3=Moderate
4=Loud
5=Very loud
6=Heard with stethoscope just off chest wall

Does the murmur occur at diastole or systole?

Determine the pitch, is it high, medium or low?

Determine the quality: blowing, rumbling, harsh or musical?

Blood pressure (BP) measurement

When assessing B/P keep the cuff about 2.5 cm above the antecubital space and use the bell for sounds low in pitch. Take the BP on each arm in lying, sitting and standing positions.

CARDIOVASCULAR PHYSICAL ASSESSMENT FORM

Inspection

General appearance/posture ______________________________

General skin color __________ Cyanosis: Mouth __________ Nail beds __________

Clubbing of fingers __________ Skin ulcerations __________

Chest symmetrical __________ Visual pulsations/location __________

Varicose veins ______________________________

Palpation

Apical impulse __________ Cardiac thrills/location __________

Radial pulse __________ Femoral pulse __________ Popliteal pulse __________

Dorsalis pedis pulse __________ Carotid pulse __________ Homan's sign __________

Auscultation

Apical pulse __________ Rhythm __________ Intensity __________

S3 __________ S4 __________ Apical/radial __________

Aortic valve (2nd intercostal space RT) S2 S1 __________ S1=Lubb

Pulmonic valve (2nd intercostal space LT) S2 S1 __________ S2=Dubb

Tricuspid valve (lower LT sternal border at 5th) S1 S2 __________

Mitral valve (5th intercostal space) S1 S2 __________

Murmurs __________ Location __________ Systolic __________ Diastolic __________

Murmur intensity __________ Pitch __________ Quality __________

Rate change with inspiration __________ expiration __________

Rt. Blood pressure ____/____ Lying ____/____ Sitting ____/____ Standing

Lt. Blood pressure ____/____ Lying ____/____ Sitting ____/____ Standing

Jugular venous distention __________ Bruits carotid artery __________

Assessment notes:

____

____

____

____

____

Section III – LABORATORY & DIAGNOSTIC TESTS

Table of common laboratory tests

Test name	Indications	Comments
Aspartate Aminotransferase (AST) Cardiac enzyme Normal: 4-36 U/ml	1. Acute myocardial infarction 2. Angina pectoris 3. Liver disease 4. Suspected damage to heart muscle	**Regarding collection:** Use red-top tube collect 5-10 ml List all drugs patient is taking IM injections before testing can raise AST **Results:** Level will increase above 40 within 6-12 hours of cardiac damage Peak levels occur within 24-48 hrs. Levels return to normal 4-6 days; Increase may peak at 3-5 X's normal
Creatine Phosphokinase (CPK) or **Creatine Kinase** (CK) Cardiac enzyme Normal: Male 55-170 U/L Female 30-135 U/L **CK Isoenzymes** MB, BB, MM	1. Acute myocardial infarction 2. Angina pectoris 3. Suspected damage to heart muscle 4. Skeletal muscle disease 5. Cerebral vascular disease	**Regarding collection:** Use red-top tube collect 5 ml Note on lab slip number of IM injections in last 24-48 hours **Results:** Level will increase within 3-6 hrs of cardiac damage Peak levels within 18-24 hrs. Levels return to normal in 3-4 days CPK-CK elevated in all problems CPK-MB up in acute MI and is the most specific test for MI CPK-BB up in cerebral problems CPK-MM up in muscle disease/trauma
Lactic Dehydrogenase (LD/LDH) Cardiac enzyme Normal: 145-450 U/L	1. Acute myocardial infarction 2. Congestive heart failure 3. Anemia 4. Suspected damage to heart muscle 5. Skeletal muscle disease	**Regarding collection:** Use red-top tube collect 5 ml Do not shake tube, avoid hemolysis Note on lab slip all narcotics and IM injections within last 8 hrs. **Results:** Level increased within 6-12 hrs of cardiac muscle damage Peak levels within 1-2 days Levels return to normal 6-10 days
LDH Isoenzymes LDH1 17-27% LDH2 27-37%		In myocardial infarction LDH1 will have a greater level than LDH2. In other words, the normal ratio is flipped

Cholesterol 150-250 mg/dl 3.9-6.5 mmol/L	1. High blood pressure 2. Cardiac disease 3. Family history of high cholesterol 4. Arterial disease	**Regarding collection:** NPO for 12 hrs, check orders Use red-top tube collect 5-10 ml Note all medications on lab slip **Results:** High levels indicate need for lifestyle and dietary changes
Triglycerides Normal 10-190 mg/dl 0.11-2.09 mmol/L	1. High blood pressure 2. Arterial disease 3. Hyperlipo-proteinemia 4. Uncontrolled diabetes	**Regarding collection:** NPO for 12 hrs., check orders Use red-top tube collect 5 ml **Results:** High levels indicate need for lifestyle and dietary changes

Section IV – PROCEDURES AND CONDITIONS

Blood pressure measurement

Blood pressure (BP) is simply the amount of pressure the blood exerts on the inside wall of a vessel.

Measurement of blood pressure

Equipment needed to measure blood pressure:

Stethoscope with bell (use bell for low-pitched sounds)
Sphygmomanometer with proper size cuff
Cuff width should be about 40% of the arm circumference or about 12-14 cm in the average adult
A cuff that is to narrow will give a false-high reading
Cuff bladder length should almost encircle the arm

Proper placement of arm and cuff:

The arm should be free of clothing
The arm should be slightly flexed to promote relaxation
Place cuff on the arm with the lower border about 2.5 cm above the antecubital space
The cuff should lie at the level of the heart
A cuff placed to low will give a false-high reading
A cuff placed to high will give a false-low reading
Cuff should fit snug, if loose it can give a false-high reading
Support the cuffed arm or let it rest on a surface
If the person supports their own arm, muscular contraction will occur and the measurement will be a false-high reading

Measurement of blood pressure:

Place the manometer dial so it is level and facing you
Locate the brachial artery on the inner part of the arm, above the antecubital space and near the biceps tendon
Keep your fingers on brachial pulse and inflate the cuff to 30 mm Hg above where you feel the pulsations stop. This is the palpatory systolic pressure reading
Slowly deflate cuff feeling for the last pulsation
Wait 30 seconds
Place bell of stethoscope over brachial artery
Inflate cuff to 30 mm Hg above palpatory systolic pressure
Deflate cuff slowly, about 3 mm Hg per second
Note reading when you hear two consecutive beats. This is the systolic pressure
Continue slow deflation until no beats are heard
Note this reading, it is the diastolic pressure

Let remaining air out rapidly until reading is zero
Record the two readings so you won't forget them
Repeat on other arm. BP should be taken in both arms in the same position at least once. If a difference is found, use the arm with the higher readings in the future. A difference of over 10 mm Hg should be reported
Pressure should be taken in three positions if cardiac problems or fainting are present. Take BP in lying, sitting and standing positions

What do the readings mean?

A blood pressure greater than 140/90 is considered high. A reading greater than 140/90 on three separate occasions is high blood pressure, or hypertension. This table is often used to diagnose the degree of hypertension. Blood pressure changes slightly from minute to minute.

Diastolic blood pressure (mmHg)	Systolic Blood Pressure (mmHg)		
	Less than 140	140-159	160 or greater
Less than 85	Normal blood pressure	Borderline isolated systolic hypertension	Isolated systolic hypertension
85-89	High normal blood pressure		
90-104	Mild hypertension		
105-114	Moderate hypertension		
115 or greater	Severe hypertension		

From the National Committee on Detection, Evaluation and Treatment of High Blood Pressure, 1984 Report

High blood pressure is not in itself a disease, it is a warning. Hypertension has become a serious national health problem because a high percentage of the population suffers hypertension, and compliance is a problem. The consequences of untreated hypertension are:

Enlargement of the left ventricle of the heart
Damage to the kidneys and possible kidney failure
Damage to the eyes and possible blindness
Damage to the blood vessels with increased risk of stroke
Damage to the heart with increased risk of heart attack
Threat of earlier death

What causes high blood pressure?

The exact cause of hypertension is unknown. Most researchers believe there is no single cause. What is known is there are several risk factors.

Risk factors that cannot be changed are:

Family history of hypertension, stroke, heart failure or heart attack.

Male sex, approximately 33% of men have hypertension, approximately 27% of women have hypertension.

Race, 10-29% of white adults have hypertension while 20-38% of black adults have hypertension. Blacks also have two times more moderate hypertension and three times more cases of severe hypertension.

Advancing age; blood pressure usually increases with age. Half of the people over age 65 have hypertension.

Risk factors that can be changed:

Smoking increases heart rate and causes vasoconstriction of vessels, leading to high blood pressure.

High LDL cholesterol in the blood can form deposits on the walls of arteries which narrows or blocks them. A blocked artery in the heart will cause a heart attack. A blocked artery in the brain will lead to stroke. Narrowed arteries have smaller vessel diameters and are less elastic resulting in increased peripheral resistance. The increased resistance means it takes more pumping force from the heart for the blood to circulate.

Obesity, lack of exercise and stress: people with these characteristics have a higher incidence of heart disease.

High fat diet; linked to a buildup of fatty deposits in the blood vessels called atherosclerosis.

What are the symptoms of hypertension?

Asymptomatic -- Hypertension is called the "silent killer" because there are no symptoms until damage is done. These are common symptoms:

- Headache -- Headaches in the morning
- Visual disturbances and reports of dizziness
- Flushing of the face
- Fatigue
- Nose bleeds
- Nervousness

Treatment of hypertension

The treatment of hypertension is lifelong; there is no "cure," but there are ways to lower blood pressure. These include:

Diet changes and weight loss
Coping techniques to reduce stress
Exercise on a regular basis
Medications
Regular blood pressure check-ups

Diet changes and weight loss

A low sodium (salt) diet is usually recommended.

- Sodium, when taken in quantities larger than 6 gram/daily, and not excreted by the body can cause fluid retention. The extra fluid increases blood volume in the vessels and increases the pressure leading to hypertension
- A low-calorie, low-fat diet may be recommended
- Obesity can also increase blood pressure, and excess weight should be reduced with the supervision of a physician

Coping with stress

- A regular daily relaxation time is important. Deep breathing and progressive relaxation techniques can be helpful
- It is important to understand that while stress and tension may cause the blood pressure to rise for a temporary period, they do not cause hypertension

Regular exercise

- Check with physician before starting any exercise program
- Most effective for cardiovascular fitness when done on a regular basis, at least three times per week for 20-30 minutes
- Begin slowly and increase program as tolerated
- Exercise increases oxygen use and has been associated with the permanent lowering of diastolic blood pressure

Medications

Often a "stepped-care protocol" is recommended:

Step 1 -- Mild hypertension

- Thiazide-type diuretic (less than full dose) and/or beta blocker
- Diuretics are given to help reduce the fluid volume the heart has to pump. One side effect is a loss of potassium
- Potassium supplements are given only if indicated
- Beta-blockers reduce blood pressure by decreasing heart rate and cardiac output. Beta-blockers are not recommended for persons with congestive heart failure or peripheral vascular disease

Step 2 -- Mild to moderate hypertension

Step one medication plus adrenergic inhibiting agent or calcium channel blocker or increase thiazide diuretic to full strength.

- Central sympatholytics lower blood pressure by stimulating alpha-adrenergic receptors in the central nervous system which decrease total peripheral resistance. Usually used with diuretics and angiotensin converting enzymes (ACE) inhibitors. May be used alone, often associated with fluid retention.
- Calcium channel blockers reduce blood pressure by causing peripheral blood vessels to dilate. May be used alone or with diuretics. Also used for angina and arrhythmias.

Step 3 -- Moderate Hypertension

Add vasodilator or calcium channel blocker or captopril (ACE inhibitor)

- Vasodilators are given to relax the muscle around the blood vessels, thereby lowering blood pressure
- ACE inhibitors may be the first medication prescribed or used in Step 3 of a stepped program. Prevents angiotensin I from conversion to angiotensin II

Step 4 -- Severe Hypertension

Add Guanethidine or Captopril.

- Guanethidine is an adrenergic blocking agent that produces a gradual drop in blood pressure and heart rate. Causes dilatation and decreased venous return. Usually given with a diuretic
- If blood pressure is controlled for 6-8 months, a reduction in dosages or return to a previous step is recommended
- Individuals taking prescription medications for hypertension should check with their physician or pharmacist before taking any over-the-counter medications for fever, colds, constipation, insomnia for possible interactions

Regular physician follow-up and blood pressure checks

Follow-up is required to:

- Determine the effectiveness of the treatment plan, assess need to adjust or change medication
- Detect problems preventing achievement of BP goal
- Assess for side effects of medications
- Assess blood pressure even if patient has no symptoms

Cardiac monitoring

Cardiac monitoring is used to detect abnormalities in the rhythm of the heart. It may be done:

As part of a complete physical examination

As part of an evaluation for chest pain
Post myocardial infarction
Prior to surgery
After beginning cardiac medications such as antiarrhythmics
There are several ways that cardiac monitoring can be done:
Holter monitor worn 24-48 hours
Electrocardiogram (ECG) in physician's office, outpatient clinic or hospital
As part of a Stress Test, combined with exercise

Monitoring with a Holter monitor system

- Three-lead monitoring system
- Electrodes are applied to the chest
  White end to right side of chest (white on right)
  Black end to left side of chest
  Green below and to one side
  Electrodes record heart's electrical activity.
- No other equipment needed. The monitor unit records
- The monitor is worn for 24-48 hours
- Patient is instructed not to bathe or shower with holter unit
- A record is kept of physical/emotional activity during test
- Activity, time of day and any symptoms are recorded
- Unusual occurrences such as chest pain are noted with time
- The diary is given to the physician for evaluation
- The ECG tracing record is matched with the diary and accounts of pain are compared with concurrent ECG tracing
- Entire ECG tracing is assessed for abnormal rhythms

Electrocardiogram monitoring

- 12-lead monitoring system
  6 leads are attached to the chest
  6 leads are attached to the extremities (arms and legs)
- The patient lies supine
- The heart's electrical activity is recorded on a galvanometer
- The ECG tracing is assessed for abnormal dysrhythmia

Reading basic rhythm strips

The electrocardiogram is a record of the heart's electrical activity and consists of waves known as P, Q, R, S, T and are shown on Figure 1E.

Basic interpretation of the rhythm strip

How to read ECG strips:

- Each large box equals 0.2 seconds and has 5 smaller boxes

5 large boxes equal 1 second
15 large boxes equal 3 seconds (every 3 seconds is a divider)
30 large boxes equal 6 seconds
A 6-second strip is used to determine rate and rhythm

- Each small box equals 0.04 seconds

Identification of a normal heart beat tracing

The waves of a normal ECG represent:

P wave represents depolarization of the atrial muscle tissue
QRS complex represents depolarization of the ventricles
T wave represents repolarization of the ventricles

1. Determine the PR interval

Should be greater than 0.11 seconds (3 small boxes)
Should be less than 0.20 seconds (1 large box)
No P wave or flattened P wave in hyperkalemia

2. Determine the QRS interval

Should be less than 0.12 seconds (3 small boxes)
Prolonged wave may occur in quinidine and procainamide therapy or in hyperkalemia

3. Determine the QT interval

Should be less than 0.43 seconds usually increases with decreasing heart rate

4. Determine the rate

Count the number of QRS cycles in a 6-second rhythm strip
Multiply the count by 10 and this equals approximate heart rate
Less than 60 is bradycardia; more than 100 is tachycardia

5. Check the T wave

It may be inverted in infarction or ischemia
Narrow, peaked T wave in hyperkalemia

The ECG strip of a normal heart beat.

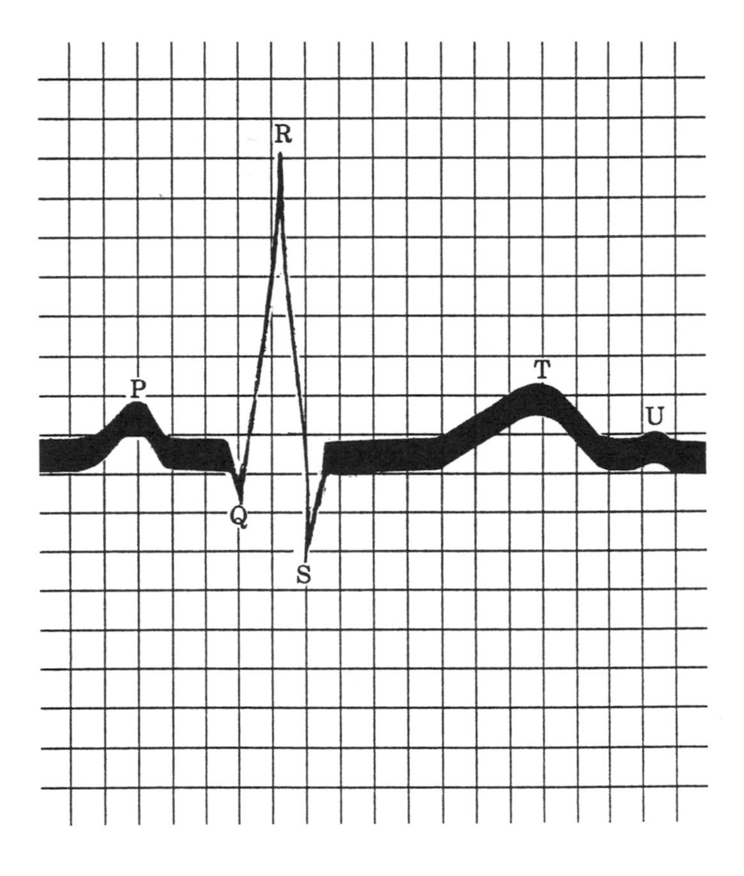

Figure 1E

TABLE OF RHYTHMS AND SIGNIFICANCE

Rhythm	Possible cause
SINUS RHYTHMS	
Sinus tachycardia Rate 100-180 Normal rhythm	Anxiety Fever Hypotension COPD Hyperthyroidism Drugs: Atropine Epinephrine
Sinus bradycardia Rate <60 Normal rhythm	Athlete Hypothyroidism Drugs: Digitalis Proranolol
Sinus dysrhythmia Rate increases with inspiration Decreases with expiration	Normal variation
ATRIAL DYSRHYTHMIAS	
Premature atrial contractions (PACs)	Stress Caffeine
Paroxysmal atrial tachycardia (PATs) Rate 140-250 Run of PAC's 3> P wave may not be visible	Heart disease; Can occur in normal person
Atrial fibrillation Rate 150-200 Irregular rhythm No P waves	Heart disease Pericarditis Mitral valve disease Pulmonary embolism Normal variant
Atrial flutter Atrial rate greater than 250 Ventricular rate QRS is variable Example is 2:1 flutter	Valvular disease Pulmonary emboli Ischemic heart

VENTRICULAR DYSRTHYTHMIAS

Premature ventricular Contractions (PVCs) QRS greater than 0.12 sec.	Caffeine Anxiety Heart disease Digitalis toxicity Hypoxia Hypokalemia Acidosis Anemia
Ventricular tachycardia Greater than 3 PVCs Rate 120-250 Wide QRS complex Life-threatening as may lead to V-Fib	Same as for PVCs
Ventricular fibrillation No cardiac output noted	Cardiac arrest

Management of a code arrest

Quick action is required when the heart stops to prevent permanent damage to the brain cells. Until help arrives, cardiopulmonary resuscitation (CPR) can be performed.

Cardiopulmonary resuscitation

CPR is the combination of artificial respiration and artificial circulation which, when started immediately, will provide oxygen to the body's tissues to sustain life.

To perform one person adult CPR:

- Assess the situation:
  What happened? Is it safe to provide care?
  Assess the level of consciousness:
  If person is conscious CPR is not needed, do not proceed
  If unconscious, call for help. Position patient on back with firm surface underneath.
  Open the airway by head-tilt/chin lift method
  Place one hand on victim's forehead. Place fingers of your other hand under bony part of victim's jaw and tilt head back while lifting jaw upward.
- Assess for respirations:
  Look, listen, and feel for 3-5 seconds
  If no respirations are detected give 2 full breaths
  Each breath is 1-1.5 seconds long
  Exhaled air has 16% oxygen enough to sustain life
- Check victim's carotid pulse:
  Check pulse on the side nearest you for 5-10 seconds
  If no pulse is present CPR must be initiated to prevent irreversible brain damage which may occur if CPR is not started within six minutes.
- Begin compressions:
  Kneel or stand facing the victim's chest
  Locate the sternal notch using hand nearest victim's legs
  Place your middle finger in notch and your index finger close to the middle finger
  Place heel of remaining hand on sternum next to index finger
  Remove first hand and place over hand on sternum
  Use only heel of hand, keep fingers off chest
  Position your body so your shoulders are directly over your hands
  Compress straight down with straight arms and locked elbows
  Compress 1.5-2 inches (adult) deep, at a rate of 80-100 per minute. Count out loud: one and two and three and...fifteen

- If cracking of ribs occurs, reposition hands and continue
- Give 15 compressions followed by 2 full breaths (1 full cycle). Relocate the correct hand position.
- Compressions are useless without oxygen from breaths. When you give a breath relocate the correct hand position. Recheck pulse after 4 full cycles:

  If pulse is present, stop CPR and monitor vital signs. Give rescue breathing as needed if no breathing is present. If no pulse, continue CPR until help arrives. Recheck pulse every 3-4 minutes while performing CPR.

To perform two-man CPR:

- Check the airway, same as one-man CRP
- Check breathing to determine breathlessness, same as one man CPR
- Circulation -- Determine pulselessness, compressor gets into position
- Compressions/ventilation ratio is 5-1. Ventilator ventilates after every 5 compressions; after 10 cycles, ventilator checks carotid pulse
- After completes fifth compression, call for switch. Ventilator compeltes ventilation, then switches

Cardiac arrest in the acute care setting

- Before an emergency, review agency policy on code arrest

  Know how a code arrest is managed at the agency. Know how to call in a code

  Find out the location of ambu bag and resuscitation cart

  Practice assembly and use of ambu bag

  Familiarize yourself with drugs and equipment in the resuscitation cart
- The signs of cardiac arrest are:

  Unresponsiveness

  No respirations

  No heart sounds, blood pressure or pulse

  Pallor and/or cyanosis

  Dilation of pupils

  An ECG pattern of ventricular fibrillation or asystole
- Call for help: Use emergency or call bell in patient's room.

  If no call bell is available, call out for assistance.
- Start CPR:

  Do not wait for help to arrive under any circumstances

  Use ambu bag or resuscitation mask for breaths if available

If after one minute no help has arrived, interrupt CPR and call in the code using patient's telephone according to hospital's policy. Once call is completed, continue CPR.
Do not abandon the patient.

- When help arrives:
  Have someone call in the code per agency policy if this has not been done. CPR should be continued, send second person for resuscitation cart.
- When resuscitation cart arrives continue CPR.
- Other personnel on scene should:
  Obtain bag/mask setup and connect oxygen if not already done. Breaths are provided, and oxygen administered via ambu bag until patient is intubated.
  Assess for presence and patency of intravenous lines, if no patent lines start one immediately for fluid and medication administration.
  Place cardiac monitor on the patient to record ECG activity. Heart's electrical activity is monitored continuously.
- Resuscitation efforts are continued until a physician orders otherwise or the patient has been resuscitated and responds with a spontaneous pulse.

Equipment needed for cardiac arrest

Most units have a crash cart and carts should be checked daily.
In general, these carts contain:

Airways in various sizes
Bag/mask setup also known as ambu bag
Oxygen flow meter and tubing
Intravenous catheters, tubing and solutions
Intravenous cutdown set
Laryngoscope and endotracheal tubes
Tracheostomy set
ECG machine (usually part of defibrillator)
Suction wall unit and tubing
Defibrillator
Variety of cardiac medications
Syringes, needles, gloves and tapes

Members of the code arrest team

The members of the code team and their function varies with the agency. All personnel should know the agency's procedure for a code arrest. Generally, these personnel respond to a code:

Physician(s) on call or emergency room physician(s)

Orders medications and fluids to restore circulation.
Monitors cardiac pattern and orders/performs defibrillation.
Coordinates and manages treatment of the arrest.
May intubate the patient.
Orders the code team to stop resuscitation if indicated.

Anesthesiologist

Intubates the patient to maintain airway.
May place intravenous line if not already done.
Ventilates by ambu bag or endotracheal (ET) tube.

Respiratory therapist

Draws arterial blood for blood gases assessment.
Provides oxygen by bag/mask or ET tube.
Suctions patient to maintain airway.

Nurses (May have from 2-4 present)

First Nurse

Is usually the first responder on the scene. Provides ventilation until others arrive. Performs CPR compressions throughout code.

Second Nurse

Starts, monitors, and maintains all intravenous lines. Places leads for ECG monitoring, connects monitor. Gives all intravenous medications as ordered. Calls out name and dosage of each medication as given. May defibrillate (if ACLS certified).

Third Nurse

Records all events of code including:

Type of arrest-respiratory or cardiac
Was it witnessed and by whom?
When resuscitation effort was begun, by whom and how
Time code was called
Name of physician who responded and time
Type of artificial ventilation used (ambu bag, ET)
Time of intubation, size of tube and by whom
All vital sign readings and time done
Time of defibrillation and watts/second
Medications given: name, dosage, route, time.
Time spontaneous heart beat and breathing began.
Time consciousness was restored.

Time resuscitation efforts were stopped.
Collects all ECG strips and lab slips.

Fourth Nurse

May defibrillate (if ACLS certified), provide ventilations, prepare medications/solutions, and monitor ECG.

Remember:

- During defibrillation stand clear of the patient and bed to prevent electrical shock.
- When defibrillating be sure the synchronizer switch is in the off position.
- After defibrillating check the monitor for rhythm pattern.
- Use defibrillating pads, saline soaked gauze or electrode gel to prevent burns.
- Discard needles in proper container to avoid needle sticks to patient and hospital personnel.

Starting an intravenous infusion

Steps in starting an intravenous infusion

Equipment:

Tourniquet
Antiseptic
Catheter -- Angiocath or butterfly in appropriate size
Gloves
Tape to secure site
Dressing (Op site or gauze)
Ordered IV solution
Tubing
IV pump or stand

Steps:

Check physician's orders
Exercise Universal Precautions
Flush tubing
Select insertion site: When possible use non-dominant hand or lower arm, assess all veins before attempting insertion
Put on gloves
Cleanse site well and allow to dry
Apply tourniquet 4-6 inches above site
Insert needle 1 cm below vein at 20-45 degree angle
Advance catheter into vein, insert up to 1/4 inch
Watch for blood return
Advance only the catheter while maintaining position of needle
Advance needle if using butterfly
Release tourniquet
Remove needle and connect tubing to catheter
Open up tubing and start pump
Observe site for swelling, burning or redness
Remove catheter if swelling or pain occurs
Tape securely and apply dressing

Common problems and solutions:

- Tape sticks to gloves: Tape the fingertips of the gloved hand before starting.
- Poor venous distention: Lower the extremity to increase venous distention. Have patient contract and relax hand. Strike the vein or apply a warm pack.
- Veins roll: Hold vein in place with free hand once skin is punctured. Enter from the top of the vein not the side.

- Small, fragile veins: Use a butterfly needle instead of an angiocath, or try a smaller gauge angiocath such as a #24.
- Need blood sample at the time of IV insertion: Once blood return is noted connect syringe to catheter hub and withdraw blood sample slowly to prevent venous collapse. Once blood sample is obtained, connect IV tubing.
- Difficulty taping the IV when the tubing is on: Place a small narrow piece of tape (sticky side up) beneath the catheter hub prior to insertion. Once tubing is connected simply fold tape over catheter mounts.
- Blood drips when trying to connect tubing: Until you have practiced enough to achieve connection quickly place your thumb over the vein just below the insertion site.
- Slow flow of solution:
  Check clamp, is it open? Is the IV patent?
  Raise the level of the IV solution. IV site may be positional, change position of extremity.

Central venous lines

A central venous line may be inserted for a variety of reasons to include:

Introduction of intravenous fluids
Administration of medications
Introduction of nutrition or hyperalimentation
Measurement of central venous pressure (CVP)
Measurement of heart's pumping ability and parameters such as Swan-Ganz catheter
Inability to obtain peripheral access

Assisting with the insertion of a central venous line:

Have the following available before procedure begins:

- Signed informed consent
- CVP tray or minor procedure tray (check agency policy)
  Antiseptic solution prep
  Gloves
  Sterile drape for sterile field
  Syringe with 25 gauge needle and anesthetic, usually lidocaine
  Suture needle and specified gauge or silk
  Occlusive dressing
- Instrument tray if not included in CVP tray
- Masks for everyone, including patient
- Gowns

- Central venous catheter, check orders for type: Intracath, Hickman, Broviac, Swan-Ganz
- Solution ordered and IV line flushed
- Heparin lock may be ordered instead of solution

Immediately prior to procedure:

- Place patient in Trendelenburg to prevent air embolism
- Place rolled towel between patient's shoulder blades to assist physician in correct catheter placement
- Instruct and assist the patient to turn their head opposite of placement site
- Place mask on patient
- Use strict sterile technique in assisting physician
- May need to draw up Lidocaine, assess patient for distress during procedure, and monitor vital signs
- Be ready with tubing and solution or heparin lock cap

After the procedure:

- Apply dressing and antiseptic ointment
- Call for chest x-ray to verify catheter placement and assess for pneumothorax
- Adjust rate of intravenous solution flow to ordered amount
- Assess for complications of CVP line and monitor frequently

Discontinuing a CVP Line:

- Remove dressing using aseptic technique
- Clamp off IV
- Remove sutures
- Withdraw catheter applying steady, gentle pressure
- Apply pressure with sterile 2 X 2 for 2-5 minutes
- Apply sterile dressing and antiseptic
- Assess periodically for bleeding or signs of infection

MAJOR COMPLICATIONS OF CENTRAL VENOUS LINES

Complication	Signs and symptoms	Prevention
Infection	Fever, chills, redness, swelling, pain, tenderness, drainage at site, Increased WBC's, Tachypnea	Aseptic technique, Change tubing q 48 hrs, Change TPN tubing q 24 hrs, Provide discharge teaching
Thrombosis	Edema of arm Pain neck/arm Jugular venous distention	Monitor site carefully Monitor PT and PTT Check for resistance to flushing, report if present
Air embolism	Confusion, anxiety pallor, tachycardia, tachypnea, hypotension, unresponsiveness	Check for leaks In catheter Tape catheter connections Restrain confused patient Position in Trendelenburg for catheter placement Keep clamp at bedside in case of accidental damage to catheter lumen Flush tubing of all air
Infiltration	No blood return, Pain or swelling at site	Always check for return leaking and swelling at site prior to infusion

Central venous pressure monitoring

Central venous pressure (CVP) measures the right ventricular filling pressure and is used to assess the ability of the right side of the heart to fill and pump blood. A CVP catheter is inserted into the subclavian vein or internal jugular. This intravenous line usually has 2 or more lumens. In order to assess the amount of central venous pressure a manometer and IV solution bag must be attached to one of the lumens.

When reading the CVP, remember:

- Zero mark on manometer should be placed at the phlebostatic axis, which is the crosspoint of line from the fourth intercostal space at a point where it joins the sternum, and a line midpoint between the anterior and posterior surface of the chest. This position should be marked for subsequent readings.
- Manometer should fluctuate with respirations
- CVP readings should be between 4-10 cm H_20
- A high CVP reading may indicate CHF, cor pulmonale, COPD, or ventricular failure
- A low CVP reading may indicate hypovolemia

- Ventilators will alter CVP readings. Alteration due to changes in intrathoracic pressure, if possible remove patient from ventilator when taking reading.
- Head of bed should be flat unless contraindicated

Section V -- DIETS

Reduced calorie diet

Indications	**Obesity**
Comments	Calories allowed will vary by need. Usually from 1200-1600 calories/day allowed. Weight loss more successful when combined with exercise. Encourage reduced portion sizes. Encourage patient to plan daily menu. Artificial sweeteners may help reduce cravings for sugar. Low calorie snacks like fresh fruits, vegetables, plain popcorn, are encouraged. Eight glasses of water daily recommended.
Restrictions	High fat foods: margarine, creams, gravies, oils. High calorie desserts and sweets: cakes, pies, ice cream, candy. Fried foods: French fries, deep fried meats.
Allowable foods	Foods from four food groups should be included. Exchange list is recommended.

Sodium restricted/Low sodium/No added salt

Indications	Retention of sodium increases BP. High BP, congestive heart failure, renal disease, or fluid retention.
Comments	Amount of Na restriction should be specified: No salt added = 4,000 mg/day Mild restriction = 2,000-3,000 mg/day Moderate restriction = 1,000 mg/day Strict restriction = 500 mg/day Severe restriction = 250 mg/day Salt free or no salt diet is impossible as sodium is found in most foods naturally. Sodium is an essential nutrient for life. Lowest rate of compliance of any prescribed diet. 1mEq of Na = 23 mg Na
Restrictions	**No salt added:** Salt if allowed in cooking, but no salt permitted after food is cooked. Foods high in sodium are limited. Processed foods are limited or eliminated: crackers, soups, canned meats, hot dogs, lunch meats, canned foods (except fruits), baking soda, baking powder, monosodium glutamate, and soy sauce. High sodium meats, cheeses, organ meats, shellfish.

Mild restriction: Limit amount of salt used during food preparation, no salt allowed at table, and omission of very salty foods.

Moderate restriction: 1/4 teaspoon salt per day allowed. No salt used in preparation of foods. Regular margarine, bread may be used; no salty foods allowed. These vegetables restricted: artichokes, beets, carrots, celery, spinach, sauerkraut, and turnips.

Foods allowed in a one gram Na diet include:
Vegetables: low-sodium canned vegetables; fresh vegetables (unless on restricted list); and frozen vegetables processed without sodium.

Fruits: all fresh allowed; and canned/frozen without sugar due to calories.

Dairy products: unsalted cottage and cheddar cheese; and whole or skim milk.

Breads and cereals: unsalted breads, low sodium crackers; unsalted cooked cereals: oatmeal, farina; unsalted dry cereals: puffed rice, wheat; unsalted pasta: macaroni, noodles, spaghetti; rice or barley prepared without salt; flour (not self rising); and unsalted popcorn.

Fats: unsalted margarine, butter, or oils; unsalted mayonnaise, salad dressing; and unsalted nuts.

Meats: fresh or frozen meats such as beef, veal, chicken, turkey, and pork not cured in salt; fresh fish (except shellfish); and dietetic canned fish.

Strict restriction: no salt in food preparation or table use; no salty foods; low sodium bread and cereal; milk restricted to 2 cups/day; meat restricted to 5-6 oz/day plus 1 egg; and vegetables listed above are restricted.

Severe restriction: Same as above restrictions. In addition meat restricted to 2-4 oz/day; eggs 3/week; and low sodium milk required.

Other Interventions

All labels should be checked for amount of sodium even when "low sodium" or "light."

Mouthcare products: toothpastes, mouthwashes may be high in sodium, rinse after use.

Medications high in sodium include aspirin, cough medicine, laxatives.
No medications should be taken without checking with physician.

Section VI -- DRUGS

The tables supply only general information. A drug handbook should be consulted for details.

ANTIARRHYTHMICS

Action: To reduce electrical irregularity of the heart; to correct dysrhythmias.

Indications: Abnormal electrical pattern noted on ECG or Holter monitor, possible irregular pulse, complaints of palpitations, weakness, dizziness. Symptoms may occur when stimulants (caffeine, cigarettes, medications) are taken, or they may occur spontaneously. There may be no obvious symptoms.

General comments: Should be taken exactly as ordered. ECG or Holter monitor is usually ordered prior to start of treatment and periodically to assess progress. Use OTC drugs cautionally. During IV therapy, monitoring of VS is important, especially the rate and character of the pulse and BP readings.

Examples of drugs in this classification:

Generic	Trade	Comment
Quinidine	Duraquin Dura-Tabs Quinatime Quin-Release	Given by PO, IM, or IV Tablets should not be crushed or chewed Continuous ECG monitoring with IV use, monitor BP (hypotension) with IV use, monitor and report the following symptoms immediately: QRS complex is prolonged No P wave is present Increase or onset of PVCs Tachycardia Drop in BP
Procainamide	Procan Promine Pronestyl	Given PO, IM, or IV Tablets may be crushed, may be mixed with food, monitor temperature during initial use, serum checks are required to monitor drug levels, pulse should be monitored daily, a record kept, and taken to physician on each visit

Phenytoin	Dilantin Novophenytoin	Also used as anticonvulsant, tablets may be crushed, may be given PO or IV Should not be discontinued suddenly Should be given with food or drink Caution against alcohol use, which increases drug blood levels, discontinue if rash appears and notify physician Balanced diet is important, regular serum drug levels required, therapeutic is 10-20 mcg/ml
Lidocaine		Given IV, for emergency use in cardiac arrest IV rate must be monitored closely Continuous ECG monitoring, convulsions may occur with IV use, seizure precautions may be helpful, Serum levels are required, therapeutic is 1.5-6 mcg/ml
Verapamil	Calan Isoptin	Given PO, IV, for emergency use In cardiac arrest Monitor I&O during initial treatment for renal impairment, pulse should be taken prior to each dose, irregular rhythms or bradycardia should be reported, dizziness during initial treatment is common, limit caffeine
Propafenone	Rhythmol	Asses for fine tremors, dizziness, Given PO
Acebutolo	Sectral	Given PO, tabbet may be crushed; reduce dose in renal disease
Esmolol	Brevibloc	Given IV. Assess for bronchospasm, rapid changes in B/P, pulse which can cause shock
Bretylium Tosylate	Bretylol	Given IV or IM. Monitor ECG continuously, use infusion pump; rebound hypertension after 1-2 hrs.
Adenosine	Adenocard	Given IV bolus. Monitor ECG intervals PR, QT, QRS

ANTIHYPERTENSIVES -- Includes several classifications; diuretics, beta-adrenergic blocking agents, angiotensin converting enzyme (ACE) inhibitors, calcium channel blockers, sympatholytics and arteriolar dilators.

Action: Reduce blood pressure.

Indications: Blood pressure greater than 140/90 on three separate occasions. There may be no noticeable symptoms or there may be complaints of headaches.

General comments: Diastolic over 120 requires immediate medical attention. When hypertension has been diagnosed BP should be taken in both arms in three positions: lying, sitting and standing. Non-compliance is a problem with these medications; it should be stressed that hypertension is a silent disease, often with no or very subtle symptoms. Medications are used with diet, weight loss and exercise to lower the BP. All patients should be encouraged to discuss side effects with their physician, as selection of the correct drug for the individual may take some trial and error. Missed doses should be taken when remembered, double doses should not be taken. A medical alert bracelet is recommended. A BP measurement should be taken prior to administration of any of these drugs, and drug withheld if the patient is hypotensive.

Diuretics, general comments: Commonly known as "water pills," these increase urine volume. Encourage fluid intake unless otherwise specified. Diuretics are usually taken in the morning. Weight, intake and output, blood urea nitrogen (BUN) and creatinine levels should be monitored.

Examples of drugs in this classification:

Generic	Trade	Comments
Bumetanide	Bumex	Given PO, IM, IV, very powerful loop diuretic, hypokalemia may occur and potassium should be included in diet daily
Chlorthalidone	Hygroton Hylidone Uridon	Given PO one time daily Caution about alcohol use
Furosemide	Lasix	Given PO, IM, IV. IV injection should be slow, 4 mg/min to prevent ototoxicity, long term exposure to sunlight should be limited, wear sunscreen, protect drug from light

Hydrochlorothiazide	Apo-Hydro Chlorzide Esidrix Hydrodiuril Thiuretic	Given PO, check orders carefully for frequency, administer after meals (PC) Obtain daily weights Also available in combination with another medication to prevent hypokalemia. Common trade names include dyazide, maxide or moduretic
Metolazone	Diulo Mykox Zaroxolyn	Given PO, administer after meals, trade medication Diulo and Mykox are not bioequivalent and should not be interchanged
Spironolactone	Aldactone	Given PO, give with food Tablet may be crushed, potassium supplementation is not needed

Beta-adrenergic blocking agents, general comments: These medications decrease heart rate and cardiac output. Before administration take apical pulse, withhold medication and notify physician if below 60. Monitor BP and assess weight daily. OTC medications should be checked with physician prior to use. Medication should not be stopped abruptly.

Examples of drugs in this classification:

Generic	Trade	Comments
Atenolol	Tenormin	Given PO, tablets may be crushed
Metoprolol	Lopressor Betaloc Norometoprol	Given PO, take with food, maximum effect seen in 1 week, avoid cold Should not be withdrawn abruptly
Nadolol	Corgard	Given PO, report irregular pulse
Pindolol	Visken	Given PO, always take same way (with or without food)
Propranolol HCl	Inderal Novopranol Propranolol	Given PO, IV, recommended before meals and bedtime Tablet may be crushed, monitor ECG, if given by IV route, medication should not be stopped abruptly

Angiotensin converting enzyme (ACE) inhibitors, general comments: Prevent Angiotensin I from conversion to Angiotensin II. Often given with diuretics. OTC medications should not be used with this medication.

Examples of drugs in this classification:

Generic	Trade	Comments
Captopril	Capoten	Given PO, close monitoring at onset of treatment is required, Administer 1 hour before meals, urine testing for protein may be ordered
Enalapril		Given PO, administer with food or liquid
Fosinopril	Monopril	Given PO. Give decreased dose in renal impairment Assess cardiac status at beginning of therapy
Quinapril	Accupril	Given PO. Give decreased dose in renal impairment; adjust according to B/P response
Remipril	Altace	Given PO Give decreased dose in renal impairment Assess for symptomatic hypotension.
Diltazem	Cardizem Cardizem SR	Given PO, AC Monitor blood level (0.025 - 0.1 mg/ml therapeutic)
Isradipine	DynaCirc	Given PO Monitor for irregular heartbeat, edema, dizziness, nausea, hypotension
Nicardipine	Cardene	Given PO Monitor for edema, dizziness, nausea, hypotension, heartbeat

Calcium channel blockers, general comments: Reduce BP by causing peripheral blood vessel dilation. Used commonly for angina and arrisythmias. Monitors BP.

Examples of drugs in this classification:

Generic	Trade	Comments
Nifedipine	Adalat Procardia	May be given PO or placed under tongue, may cause gum bleeding or redness, smoking is not recommended Check orders before surgery, medication may need to be withheld

Verapamil	Calan Isoptin	Given PO or IV, administer with food, check apical pulse before administration withhold if below 60

Centrally acting adrenergics, general comments: Lower BP by decreasing total peripheral resistance. Often used with diuretics due to associated fluid retention. BP should be monitored, orthostatic hypotension may occur. Advise patient driving is not recommended during initial therapy.

Examples of drugs in this classification:

Generic	Trade	Comments
Clonidine	Catapres Dixarit	Given PO or via transdermal patch, do not withdraw medication suddenly, dry mouth is common complaint, eye examination is recommended, place patch on clean, dry area, reinforce patch with tape Rotate patch site, assess for rash and report if found, medical alert bracelet should be worn, central sympatholytic
Guanethidine	Ismelin	Given PO, tablets may be crushed, do not take OTC Peripheral sympatholytic
Methyldopa	Aldomet Dopamet Medimet	Given PO, caution about sudden position changes as fainting may occur, central sympatholytic

Vasodilators, peripheral, general comments: Lower BP by relaxing the smooth muscle inside blood vessels causing vasodilatation. Usually given with diuretics or beta blocking agents. Orthostatic hypotension may occur with position changes.

Examples of drugs in this classification:

Generic	Trade	Comments
Hydralazine	Hydralyn Apresoline Rolazine	Given PO, instruct patient to avoid prolonged standing or exposure to heat

CARDIAC GLYCOSIDES/INOTROPIC AGENTS

Action: Increase the force of cardiac contraction while slowing the heart rate.

Indications: Congestive heart failure, cardiac dysrhythmias, cardiogenic shock from cardiac trauma (myocardial infarct).

General comments: Take apical pulse prior to administration, withhold drug if irregular or under 60/BPM. When potassium levels are low, there is increased sensitivity to drug. Therefore, potassium should be monitored. Cardiac glycocides are powerful drugs, and should be stored from children; accidental ingestion may be lethal. A medical alert bracelet should be worn. OTC medications used only with advice of physician.

Examples of drugs in this classification:

Generic	Trade	Comments
Digoxin	Lanoxin Masoxin Novodigoxin	Not the same drug as Digitoxin Given PO, IM, IV, tablet may be crushed, serum levels are required Therapeutic range is 0.8-2 ng/ml IM route is associated with intense pain Anorexia is a sign of toxicity, weigh daily, assess for edema, store in light-resistant container
Digitoxin	Purodigin	Given PO, IV, dosage is titrated Serum levels are required, therapeutic range is 20-35 ng/ml, IM route is associated with intense pain, usually taken in morning, store in light resistant container

ANTILIPEMICS

Action: To lower blood cholesterol levels.

Indications: High serum cholesterol levels that have not responded to dietary interventions.

General comments: High cholesterol levels have been associated with cardiovascular disease.

Examples of drugs used in this classification:

Generic	Trade	Comments
Cholestyramine	Questran Cholybar	Give other medications 1 hour before or 2-4 hours after this medication, dissolve in water, juice, or soup before taking Cholesterol level decreases in a month

Colestipol	Colestid	Same as Cholestyramine
Gemifibrozil	Lopid	Give 1/2 hour before meals, caution against driving or using machinery until reaction to drug is known
Lovastatin	Mevacor	Give with evening meal, blood cholesterol decreases within 2-4 weeks of therapy, avoid alcohol use
Niacin	Nicobid Vitamin B_3	Give with meals and cold liquid, Report if flushing occurs
Probucol	Lorelco	For persons who do not tolerate other antilipemics, give with meals. An ECG may be ordered at start of treatment May prolong QT interval
Simvastatin	Zocor	Given PO. Asses for symptomatic hypotension; increased mm isoenzyme of CPK

VASODILATORS, CORONARY

Action: Relaxes smooth muscle producing vasodilatation which decreases amount of blood returned to the heart and thereby decreases cardiac output.

Indications: Angina, also used with BP medications.

General comments: Headaches are common during intitial treatment, monitored BP prior to administration for postural hypotension. Medical alert bracelet should be worn.

Examples of this drug classification:

Generic	Trade	Comments
Dipyridamole	Persantine	Used for chronic angina, may be taken PO, take 1 hour before, or 2 hours after meals, tablets may be crushed.
Nitroglycerin	Nitro-bid	May be given SL, transdermal, topical or IV, used for acute angina episodes May take up to 3 tabs SL in 15 min if pain is not relieved, sustained release tablets should be taken on empty stomach When giving ointment, do not apply with bare fingers Blurred vision or dry mouth should be reported to physician

Section VII -- GLOSSARY

Aneurysm	Abnormal dilation of a vessel due to weakness in the wall
Angina pectoris	Chest pain related to insufficient blood supply to cardiac tissues. May radiate to shoulder and left arm
Angiogram	X-ray of blood vessels using contrast dye
Apex of heart	Bottom tip of the heart
Arrhythmia	Abnormal rhythm of the heart
Atherosclerosis	Build-up of fatty deposits in blood vessels
Atrioventricular node (AV node)	Part of the heart's electrical conduction system
Atria	The two upper chambers of the heart
Atrioventricular valves (AV)	The valves located between the atria and valves ventricular chambers of the heart
Bicuspid valve (Mitral valve)	Separates the left atrium and left ventricle, one of the AV valves
Blood pressure (BP)	Amount of pressure exerted on the inside wall of a blood vessel
Bradycardia	Heart rate below 60 beats per minute
Bundle of His (Atrioventricular bundle)	Part of the conduction system of the heart which conducts electrical stimulation from the AV node to the medial surfaces of the ventricles
Cardiac	Heart
Cardiac catheterization	A diagnostic procedure in which a radiopaque catheter is inserted into the heart via a vein in arm or leg and threaded into the vena cava, right atrium, or right ventricle.
Cardiac output (CO) Code	The amount of blood the left ventricle ejects per minute. Calculated: SV X HR

Code arrest	An emergency in which breathing has ceased (respiratory arrest) or the heart has stopped beating (cardiac arrest)
Crash cart	The resuscitation cart used in codes
Defibrillation	Attempting to stop the fibrillation of the heart with drugs or electrical shock; electrical defibrillation applies a countershock to the heart via electrodes on the chest
Diastole	The period of relaxation in the heart cycle when the ventricles fill with blood
Echocardiogram	Ultrasound examination of the heart, non-invasive procedure
Electrocardiogram (ECG)	A diagnostic test that records the electrical activity of the heart. Used to diagnose abnormal cardiac rhythms
Embolism	Obstruction of vessel caused by a clot or foreign substance circulating in the blood
Endocarditis	Inflammation of the endocardium
Endocardium	Thin layer of endothelium lining the inner aspect of the heart
Epicardium (Visceral pericardium)	Transparent outer layer of the heart wall
Extracorporeal circulation	Procedure to circulate blood during cardiac surgery while bypassing heart and lungs
Fibrillation	Quivering or contraction of cardiac muscle fibers, may be atrial or ventricular
Hyperkalemia	Serum potassium above level of 3.8-5.0 mmol/L
Hypertension	High blood pressure, systolic over 140; diastolic over 90
Hypokalemia	Serum potassium below level of 3.8-5.0 mmol/L

Intravenous line (IV)	A system to deliver fluids through the veins. A catheter is placed in the vein, and a flexible tubing is connected to a solution bag
Myocardial infarction (MI) (Coronary, heart attack)	Death of myocardial tissue caused by infarction in the heart
Myocardium	Cardiac muscle tissue
Palpitation	A sudden throbbing or flutter in the heart
Pericardum sac	Serous membrane sac enclosing the heart
Pericarditis	Inflammation of the pericardium
Phlebitius	Inflammation of a vein
Purkinje fibers	Part of the electrical conduction system of the heart
Semilunar valves	Two valves located between ventricles and the pulmonary artery and the aorta that prevent blood from flowing back into the heart
Sinoatrial node (SA)	The pacemaker of the heart; it initiates the cardiac cycle
Stroke volume	The amount of blood the heart ejects during contraction (systole) from the ventricle (average 70 cc/beat)
Systole	The period of ventricular contraction in the heart cycle
Tachycardia	Heart rate above 100 beats per minute
Tricuspid valve	Separates the right atrium and right ventricle so called because it contains three flaps; one of the two atrioventricular valves
Ventricles	The two lower heart chambers
Vital signs (T, P, R, BP)	Temperature, pulse, respirations and blood pressure

Respiratory System

RESPIRATORY SYSTEM

Table of contents

Section I – OVERVIEW

Primary functions

1. Respiration is the exchange of gases (oxygen and CO_2) between the atmosphere and the blood that is vital for life.

There are two types of respiration, external and internal.

External respiration is the exchange of gases between the external environment (air) and internal environment (lungs).
Internal respiration is the exchange of gases at the cellular level between the alveoli and the blood.

2. Filtering, warming, and humidification of incoming air through the nose.

3. Formation of the sounds, or speech.

Components and function

This system contains the organs used in external respiration: the nose, pharynx, larynx (voice box), trachea (windpipe), bronchi, alveoli and lungs. Each component assists in the primary function of respiration.

Nose

The nose is made of cartilage and bone, and is divided into two chambers by a septum.

The nose is the entry point for air entering the body, although the mouth can also serve this function.

The nose provides a filtering center for the incoming air. The nose and upper airway are lined with course hairs called cilia which filter dust and other particles to prevent foreign matter from entering the sterile environment of the lungs.

The cilia trap the particles in the protective mucous found inside the nose, particles are then swallowed or removed from the nose (such as when we blow our nose).

The warming of incoming air begins in the nose. The nose has a network of tiny blood vessels that warm the incoming air as it passes over them.

The nose humidifies the incoming air. The same abundance of tiny blood vessels that warms the air also aids in humidifying it by adding moisture as the air passes. The mucous found inside the nose also helps by producing about two cups of moisture each day.

Pharynx (The back portion of the nose and mouth)

The pharynx is a muscular tube which extends from the base of the skull to the level of the sixth vertebra. It is only five inches long and has three parts: the nasopharynx, the oropharynx, and the laryngopharynx.

On one end of the pharynx is the nose; on the other end it communicates with the esophagus and the larynx.

The pharynx continues the air passage begun in the nose. Once the air enters the body through either the nose (preferred route) or the mouth (secondary route), it travels down to the pharynx where it continues to be humidified and warmed.

Besides the part the pharynx plays in respiration, it also provides the passageway for food from the mouth into the esophagus and the stomach.

Larynx (also known as the voice box)

The larynx is the link between the pharynx and the trachea. The larynx produces the protrusion in your neck commonly called the Adam's apple.

The adult larynx is approximately 36 mm long in females and 44 mm in males. It consists of nine cartilages bound together by an elastic membrane and muscle.

The epiglottis is a thin structure (flap) that covers the entrance to the larynx when swallowing occurs; if the larynx was not covered food or drink would enter the lungs and aspiration would occur. If an object does enter the larynx by accident, the airway will close off; this is followed by a cough to expel the foreign substance.

It is in the interior of the larynx that the ventricular folds (false vocal cords) and the vocal folds (true vocal cords) are found.

The larynx like the nose and pharynx is also coated with mucous and the process of filtering incoming air is continued.

Trachea (also known as the windpipe)

The trachea is a large tube made up of bands of C-shaped cartilage and is about 4 1/2 inches in length.

The trachea is located between the larynx and the bronchial tubes. The beginning of the trachea is found at the sixth cervical vertebra and extends to the fifth thoracic vertebra.

The trachea divides into the right and left bronchi, the point of division is the carina tracheae.

The trachea provides a mucous-lined passage for air movement into and out of the lungs.

Bronchi (Bronchus = Singular)

The bronchi are located inside the lungs.

The right bronchus is shorter, wider and more vertical than the left bronchus; for this reason foreign objects are more likely to become lodged in the right bronchus than the left.

Both bronchi branch out like a tree; there are five smaller bronchi which lead into the five lobes of the lungs. These smaller bronchi further divide into about 50-80 bronchioles in each lobe.

The terminal bronchioles further divide into two respiratory bronchioles. Each of the two respiratory bronchioles divide further into alveolar ducts.

Alveolar ducts give rise to alveolar sacs and alveoli (about 300 million in the lungs), where internal respiration, or gas exchange, takes place.

Lungs

There are two lungs which are connected to the trachea by the right and left bronchi. The right lung is thicker and broader than the left and has three lobes, the left lung has only two lobes and an indentation for the heart known as the cardiac depression.

The lungs contain the millions of alveoli where gases are exchanged with the blood. The total surface area for air exchange in the average adult is 50-100 square meters.

The purpose of the lungs is to bring air and blood into close proximity. O_2 and CO_2 can be exchanged in internal respiration.

Concepts

Breathing

The average healthy adult at rest breathes 14-20 times/min. With each breath about 500 ml of air is inspired. The average maximum amount of air that can be inhaled is about 3000 ml and is called the inspiratory reserve.

The maximum amount of air that can be exhaled after a normal breath is about 1200 ml and this is called the expiratory reserve. The amount of air that can be exhaled forcibly after inhaling the largest breath possible is about 4700 ml, and this is called vital capacity.

The amount of air left in the lungs that is not exhaled, even forcibly is about 1200 ml and is known as the residual volume.

Process of internal respiration

Oxygen deficient blood enters the lobes of the lungs (three on the right, two on the left) through the pulmonary arteries leaving the heart.

The pulmonary arteries are large and then progressively branch and narrow, until they become capillaries.

Capillaries, connect the arterioles (smallest vessel in the arterial system) with the venules (smallest vessels in venous system).

Capillaries, like the alveoli, are located throughout the internal surface of the lungs. Once the blood has traveled from the heart via the pulmonary arteries into a capillary, it is exposed to oxygen on all sides as it passes through the alveolar wall.

Each individual blood cell has millions of hemoglobin molecules in it; each hemoglobin molecule is able to bind four molecules (eight atoms) of oxygen. At the same time the hemoglobin molecule is picking up oxygen; it is releasing carbon dioxide molecules. This exchange of gas molecules is internal respiration.

Control for respiration

Unlike the heart, the lungs do not expand and contract spontaneously. Information from the central nervous system controls the diaphragm and intercostal muscles needed for respiration.

The respiratory center located in the lower portion of the brain stem (medulla oblongata) receives constant information from the body, and this allows the respiratory center to change the rate or depth of inspiration or expiration, thereby controlling the amount of oxygen the body receives.

The central components of the regulatory system are located in the medulla; the medulla has chemoreceptors that monitor changes in the blood level of carbon dioxide.

When the level of carbon dioxide is elevated, neurons in the respiratory center increase the rate of respirations. An increase in respirations allows for more carbon dioxide to leave the body, and more oxygen to enter, thus, the balance is restored.

The increase in the rate of breathing is not a conscious act. The respiratory regulatory center sends the message to the muscles that control breathing without the person becoming aware that the carbon dioxide level has increased. If we hold our breath until we faint, the respiratory regulatory center will take over as it is stimulated by the increase in carbon dioxide levels.

Carbon dioxide levels in a healthy person will always provide the stimulus for respiration (in persons with chronic lung disease who have elevated levels of carbon dioxide, this is not the case).

How is breathing accomplished?

The brain provides the stimulus for respiration, but how does it initiate the mechanics of breathing? The act of breathing obeys the laws of physics and is simply a response to changing pressures. In between respiration, the pressure outside the body and the pressure inside the body are the same.

Once the respiratory regulatory center in the brain sends the message to the diaphragm and intercostal muscle to contract, the volume of the lungs changes. The lungs increase in depth and circumference, and this causes the pressure inside the lungs to drop below atmospheric pressure. Once the pressure drops, air moves into the lungs (inspiration); the pressure in the lungs again equals the air pressure outside the body. A message is then sent to the diaphragm and intercostal muscles to relax. This reduces the volume of the lungs, and the pressure increases above that of the atmosphere. Air is then forced out of the lungs into the external environment (expiration). (See the chart, Mechanism of Breathing.)

MECHANISM OF BREATHING

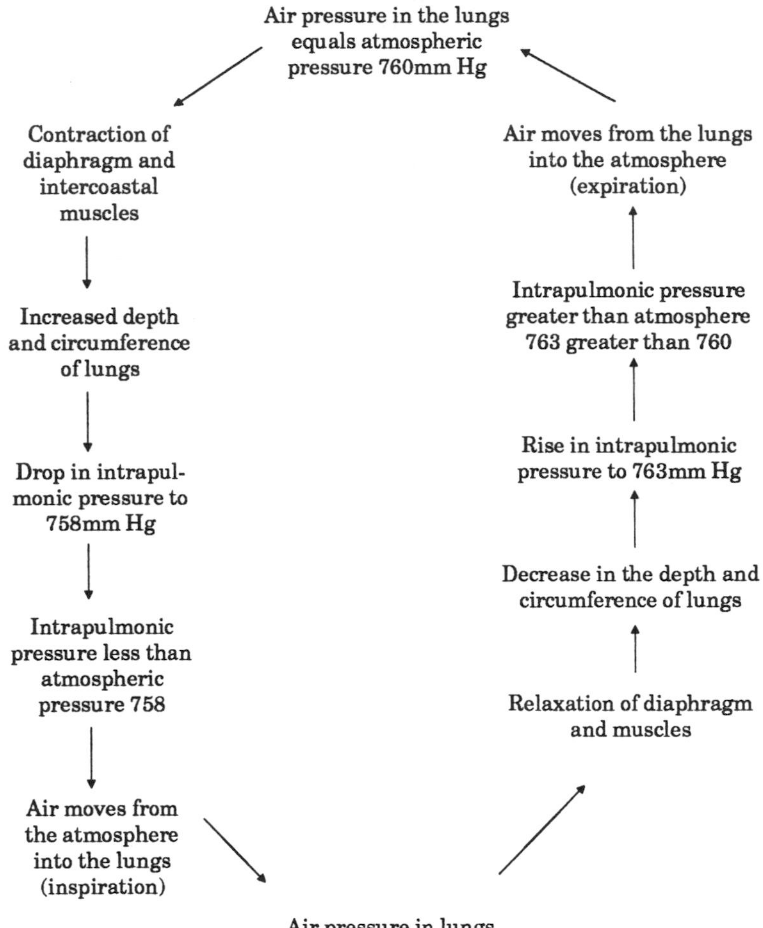

Figure 2A

Section II -- ASSESSMENT

Assessment of the Respiratory System

Assessment of the respiratory system includes the chief complaint, health history, and performing a physical assessment.

Health history

Chief complaint

In any health assessment it is important to determine what caused the patient to seek help. The chief complaint should be stated in the patient's own words when possible. The nurse assesses onset, symptoms, frequency, and duration of problem. This information will narrow the possible causes of the chief complaint and give a clinical picture of what is happening. Use simple phrases when asking questions such as:

"When did you first notice the cough?"
"How often are you coughing?"
"Is it worse at night?"
"How long does the coughing last?"

It is also important to ask what the patient has done to correct the problem before seeking health care. Have they taken over-the-counter medications? Antibiotics? Used home cures?

Examples of common chief complaints include:

Cough, productive or non-productive
Dyspnea (difficulty breathing)
Chest pain
Wheeze
Hemoptysis
Shortness of breath
Personal and family history

The history may be important in determining the patient's problem. Is there a personal or family history of chronic disease?

Health history

The history may be important in determining the patient's problem. It there a personal or family history or chronic lung disease?

Tuberculosis

For family or personal history of tuberculosis, assess date, treatments, or prophylactic medication given such as Isoniazid. Note results and date of last TB test or chest x-ray.

Lung cancer

If personal history is positive note treatment, surgery, and dates.

COPD

Chronic Obstructive Pulmonary Disease includes chronic bronchitis, asthma, and emphysema. For asthma note whether childhood onset or adult.

Shortness of breath, cough, pain in chest

The list of signs and symptoms found in this section can be helpful in assessment.

URI

URI is upper respiratory infection. Note if patient has been diagnosed with frequent upper respiratory infections or currently has an upper respiratory infection.

Productive cough

Productive cough refers to patient's ability to produce sputum.

Respiratory system testing

Pulmonary function studies

For pulmonary function studies, it may be helpful to ask if any special pulmonary tests have been done. Pulmonary function studies include:

> Ventilatory Function Tests
> Tidal Volume, Vital Capacity, Expiratory/Inspiratory reserve, Forced Inspiratory Volume, Residual Volume, Functional Residual Capacity, Maximal Voluntary Ventilation, Total Lung Capacity, and Flow Volume Loop

Other Tests

Arterial blood gases	Bronchoscopy	
Bronchogram	Chest	Imaging
Lung	Biopsy	Mediastinoscopy
Sputum culture	Thoracentesis	

Current information

List all physicians who have seen the patient and all medications taken. Careful attention should be given to medication side effects that could contribute to the chief complaint.

Cigarettes and alcohol consumption

The Surgeon General has issued a warning that cigarette smoke contains carbon monoxide. Cigarette smoking has been linked to lung and heart disease. If your patient smokes, determine if the cigarette is low or high in tar.

Excessive alcohol consumption can lead to chronic disease and it is important to determine if the patient is a light, moderate or heavy drinker.

RESPIRATORY SYSTEM ASSESSMENT FORM

Chief Complaint

Patient's statement________________ Onset________ Symptoms______________

Frequency______________ Duration______________ Other areas affected__________

Have you had this before?________ When?______________________________

What have you done for this?______________________________________

What do you think caused this to happen?____________________________

What changes have you had to make in your life because of this problem?____________

____

What does this illness mean to you?__________________________________

Personal and Family History

	Patient Date	Family Member		Patient Only Now	Patient Only In Past
Tuberculosis	____	____	Shortness of Breath	____	
Lung cancer	____	____	Wheezing	____	____
Pneumonia	____	____	Bloody sputum	____	____
COPD	____	____	Pain in chest	____	____
Asthma	____	____	Where? When?	____ ____	
Emphysema	____	____	URI	____	____

Do you have a cough?__________ Productive_________ Color of sputum____________

Respiratory System Testing

Chest X-rays____________ Pulmonary function studies_________________________

Other____

Current History

Physician_____________________________ Phone_________________________

Cigarettes/day_______/_______/______yrs Alcoholic drinks_______/______/______yrs

What is your usual weight?_____________ Have you lost/gained?__________________

Medications

Name______________________________ Dose_____ Frequency_______ Route______

Name______________________________ Dose______ Frequency_______ Route______

Physical Assessment

Physical assessment of the respiratory system is done in the following order:

1. Inspection
2. Palpation
3. Percussion
4. Auscultation

Inspection

General appearance

What position is the patient in? Some patients are not able to tolerate the supine position if they are in respiratory distress. They may position themselves in high Fowlers in order to breathe comfortably. Look carefully at your patient before you ask them to make a position change.
How does the patient look? Patients with chronic obstructive pulmonary disease (COPD) may have a cachexic appearance as the work of breathing is increased leading to fatigue and a greater caloric need.

General Skin Color/Mucous membranes/Nailbeds

Is the patient pale, flushed or cyanotic? To assess for cyanosis look at the lips, mouth and under the tongue for gray or blushish discoloration. Also look at the nailbeds for cyanosis. Cyanosis can be caused by decreased oxygen due to pulmonary or heart disease or it can be caused by the patient being cold.

Clubbing

Clubbing of fingers and curvature of nails on fingers/toes is abnormal and is usually related to chronic lung or heart disease.

Breathing: rhythm, rate and effort

Observe the rhythm, rate and effort of breathing. Normal breathing is regular, at a rate of 10-20 times per minute and unlabored. Labored, rattling breaths, wheezing or shallow respirations are abnormal for the adult at rest. In labored breathing observe for use of accessory muscles in the neck and retractions.

Symmetrical movement of chest

Compare one side of the chest to the other; are the two sides equal during inspiration?

Expirations greater than inspirations

During normal breathing at rest the expiratory phase is longer than the inspiratory phase. If the expiratory phase is excessively long, check for use of abdominal muscles aiding in respiration.

Mouth breathing/Pursed lips/Flared nostrils/Retractions/Use of Accessory muscles

All of these are abnormal in the normal adult at rest and are often used as a compensating mechanism to help bring in more oxygen. Pursed lip breathing is often associated with COPD, and flaring nostrils may be present in asthma or respiratory distress. Use of accessory muscles occurs in labored breathing; muscles involved may include the trapezius, scalenus, sternomastoid during inspiration, as well as the abdominal muscles during expiration.

Apneic Episodes

Apnea is a temporary stop in breathing. Apnea may occur from low levels of carbon dioxide in the blood, damage to the central nervous system or even be voluntary (breath holding). If apnea is noted, how many seconds did it last? How many episodes have occurred?

Cough/Sputum

Is cough present? Is cough productive? If so, is there scant, moderate or copious amounts of sputum produced? Is the sputum thick or thin in consistency? What color is it? Is there an odor that is noticeable?

Shape of chest

Is the chest shape abnormal? In barrel chest, the chest appears round, the ribs are more horizontal than normal and the sternum is prominent. Barrel chest is seen in pulmonary disease. In funnel chest, part or all of the sternum is depressed; if the depression is severe enough it can interfere with respirations. In pigeon chest, the sternum protrudes anteriorly and the chest resembles the thorax of a fowl.

Palpation

Respiratory excursion

Respiratory excursion is tested by placing examiner's thumbs on either side of the spinal cord at the level of the tenth rib with palms flat on the posterior chest. Ask the patient to inhale, observe expansion of patient's lungs; normal results are symmetrical expansion. The distance of expansion can be estimated or measured. A lack of symmetry may mean underlying disease.

Pain on Palpation

One of the purposes of palpation is to identify tenderness. If tenderness is noted, document area; use light touch over any of these areas. Also palpate masses or lesions and document their location.

Percussion

Percussion over the chest wall will assess for solid, fluid, or air filled spaces.

Resonance - hollow sound heard over most of the normal lung.

Hyperresonance - low-pitched, booming sound, abnormal; heard when lung is overinflated as in COPD (emphysema).

Tympany - loud high pitch sound (drum like); abnormal over chest, normal on stomach.

Dullness - Soft, high pitched dull; heard over heart or solid surface such as in consolidated lungs.

Auscultation

Use the diaphragm of the stethoscope to listen to your patient's lung sounds. Ask your patient to breathe deeply through mouth. Listen to one full breath at each location.

Breath sounds

Bronchial breath sounds are normal over the trachea; these are loud and high pitched BLOWING sounds. Inspiratory time should be less than expiratory time. They are abnormal when heard in other areas of the chest.

Bronchovesicular breath sounds are normal over both of the bronchi; these are medium intensity sounds. The inspiration and expiration of a breath should be equal.

Vesicular sounds are soft, light breath sounds; normal over the lung surface. In these areas, the inspiratory time is greater than the expiratory time.

Decreased/Crackles/Gurgles/Wheezes/Rubs

All of these are terms used to describe abnormal breath sounds. In crackles, the crackling, popping sound may be fine, medium or coarse in intensity. Crackles result from fluid in airways or from small areas of atelectasis that expand with deep breathing.

Gurgles are loud, coarse, low-pitched sounds somewhat like a snoring sound, that result from thick secretions in the larger airways (usually heard on expiration).

Wheezes are high-pitched, like a whistle or musical sound resulting from a narrowed airway (as in asthma); heard on expiration, but may be heard on inspiration if severe.

Plural friction rub is a rough grating sound resulting from the visceral and plural linings of the lung rubbing together.

RESPIRATORY SYSTEM ASSESSMENT FORM

Inspection

General Appearance/Posture____

General skin color________Mucous membranes________Nails_______Clubbing_______

Breathing: Regular____________Irregular_________Rate__________Unlabored_______

Symmetrical movement of chest___________Expirations/Inspirations________________

Labored_________Crackles___________Wheezing_________Shallow____Other______

Mouth breathing/Pursed lips________Flaring nostrils__________Retractions__________

Where?________Use of accessory muscles: neck_____shoulder______abdomen________

Apneic episodes________How long?______How many?____/_____Monitor on/off______

Cough______Sputum______Amount______Consistency_________Color_____Odor___

Shape of chest:________________________________Symmetrical__________________

Palpation

Respiratory excursion: Symmetrical______________Inspiration/cm's________________

Pain on palpation_____________Masses_________________Location_______________

Percussion

Resonance_________Tympany___________Dullness_________Sound symmetry_____

Auscultation

Symmetrical breath sounds: Anterior________Posterior___________Absent__________

Bronchial(Trachea/I):______________Bronchovesicular(Bronchi/I=E):________________

Vesicular(Peripheral areas/*IE):____________________Decreased___________________

Crackles___________Gurgles_______________Wheezes____________Rubs__________

Assessment Notes____

____

____

____

*I=Inspiration E=Expiration

Section III -- DIAGNOSTIC AND LABORATORY TESTS

The following table contains some of the more common laboratory tests used in this system, values are in SI units.

Name of test/ Normal value	Indications	Comments
Acid-Fast Bacilli (AFB) No Mycobacterium tuberculi found	Tuberculosis	Sputum is required, Takes 72 hours
Arterial Blood Gases Decreased pH <7.35 Increased CO_2 >35-45 mmHg HCO^3-- within normal limits 22-26 mEq/L	To determine the acid/base balance of blood COPD Emphysema Asthma Bronchitis Pneumonia Guillain-Barre *ARDS, IRDS	Blood is drawn from artery, Apply pressure to site for 5-10 minutes Respiratory acidosis Signs and symptoms include: Hypoventilation Dyspnea Disorientation Drowsiness, headache
Increased pH >7.45 Decreased CO_2 <35-45mmHg HCO^3-- within normal limits 22-26mEq/L	Pulmonary embolism Fever Anxiety	Respiratory alkalosis Signs and symptoms: Hyperventilation Anxiety Numbness, tingling Late hypoventilation convulsion
Sputum Culture	Respiratory infection Pneumonia Lung carcinoma	Sputum sample is required
Sputum for cytology No evidence of cancer cells	Lung carcinoma	Sputum sample is required
Theophylline Level 28-112 umol/L therapeutic level 112 umol/L is toxic	Theophylline drug use	Blood sample required, Theophylline is used to treat asthma, bronchoconstriction 5-20 ug/ml

This table provides information on common diagnostic tests used in the respiratory system.

Name of test	Indications	Comments
Bronchoscopy	Visualize bronchial tree, Obtain a biopsy	Consent form needed, Premedication may be given, Local or general anesthesia, Sore throat may follow exam, Post-procedure: Monitor for complications Bronchospasm Pneumothorax Bleeding Frequent vital signs Place in semi-fowlers position No liquids until gag reflex is present, usually 2-8 hrs
Bronchogram	To detect tumors obstructions, Bronchiectasis	Consent form needed, NPO for 6-8 hours before test, Assess for allergy to iodine, Expectorants may be given for several days prior to test, Premedication is given to promote relaxation, Local anesthetic may be used, An iodine contrast is used to coat the bronchi, X-rays are taken during procedure Post-Procedure: Monitor for complications Bronchospasm Bleeding Hoarseness Frequent vital signs Assess lung sounds; report any change Place in semi-Fowlers position No liquids until gag reflex is present 2-8 hrs
Chest Imaging MRI or CT scan	Trauma, detect changes in lung tissue	May be used with or without contrast dye material, Assess for allergy to iodine
Chest X-Ray	Pneumonia Tumors Lung abscess Pneumothorax	Very common procedure Takes about 10-15 minutes Contraindicated when pregnant

Mediastinoscopy	Lung metastasis Lung carcinoma	Consent form needed, NPO 8-12 hours before test Pre-operative medication is given, A mediastinoscope is inserted through incision at suprasternal notch and biopsy is obtained Post-procedure: Usual agency post-op policies Assess for decreased or absent breath sounds
Thoracentesis	Fluid in lung cavity	A consent form needed, Local anesthetic may be given, Usually performed in patient's room, patient sits, leaning over bedside table, fluid from lung(s) is withdrawn with needle, Monitor for dyspnea, chest pain, tachycardia, cough, cyanosis, dizziness, collect fluid drained and assess amount, color, consistency, monitor breath sounds and vital signs frequently

Pulmonary function tests

Pulmonary function tests are used as an aid in the diagnosis of respiratory disorders such as emphysema and asthma. With these tests, baseline pulmonary functioning can be assessed. These tests may be ordered for any patient suffering dyspnea.

Preparation for pulmonary function tests

- Assess and report any upper respiratory infections; if present, testing may be postponed. The primary focus of testing is to assess the usual baseline functioning of the lungs or to aid in the diagnosis of chronic pulmonary problems.
- Smoking should be avoided 4-6 hours prior to the test.
- IPPB (Intermittent Positive Pressure Breathing) treatments should be held 2-4 hours prior to testing.
- Avoid heavy eating or drinking which may affect the ability of the patient to perform.
- Dentures, if normally worn, should not be removed.
- The ability of the patient to cooperate in taking deep and/or fast breaths is vital to pulmonary function testing. An explanation of this is important prior to testing to increase compliance. This also allows the patient time to practice before testing.
- Record the patient's weight, sex, age and race on pulmonary function slip.

The normal expected results of pulmonary function testing are shown on the following chart. Normal values vary with age (decreased in advancing age) and sex (women have generally lower values by as much as 25 percent). Race and body weight may also affect values.

EXPECTED TEST RESULTS

Name of test	Normal Results Definition
Tidal volume (TV,Vt)	500 ml. Amount of air inhaled and exhaled in a single breath at rest. A spirometer may be used for 1 minute and total divided by number of breaths taken in the minute.
Inspiratory reserve volume (IRV)	3000-3100 ml. Maximum amount of air that can be inhaled after a normal inspiration.
Inspiratory capacity (IC)	3500-3600 ml. The total amount of air that can be inhaled in one breath. IC = TV + IRV 3600 - 500 = 3100 3600 = 3600
Functional residual capacity (FRC)	2300-2400 ml. The amount of air left in the lung at the end of a normal exhalation.
Expiratory reserve volume (ERV)	1100-1200 ml. The total amount of air that can be exhaled after a normal exhalation.
Residual volume (RV)	1200-1300 ml. The amount of air left in the lungs that you cannot exhale; this air keeps the alveoli slightly inflated. Increased RV may result from emphysema where the lungs contain an abnormally large amount of air even after maximum exhalation. RV = FRC - ERV 1200 = 2400 - 1200 FRC = ERV + RV 2400 = 1200 + 1200
Vital capacity (VC)	4600-4800 ml. Amount of air exhaled slowly and totally, after inhaling the largest breath possible. A decreased VC may result from pulmonary edema, pneumonia, collapsed lung tissue, or any disorder that inhibits full lung expansion. VC = ERV + TV + IRV 4800 = 1200 + 500 + 3100 4800 = 4800
Total lung capacity (TLC)	5800-6000 ml. The total volume of air in the lung after a maximum inhalation, this must be calculated.

	A decreased TLC may result from pulmonary edema, collapsed areas of lung, tumors, pneumonia or any condition that inhibits lung expansion. An increased TLC may result from emphysema where the lung hyperinflates. TLC = RV + ERV + TV + IRV 6000 = 1200 + 1200 + 500 + 3100 6000 = 6000
Forced vital capacity (FVC)	4600-4800 ml. The amount of air that can be forcefully exhaled quickly, after maximum inhalation, FVC usually equals VC. The measurement of FVC can be further broken down into Forced Expiratory Volume Timed.
Forced expiratory volume timed (FEVt)	Measured during the FVC, the amount of air the patient can exhale in seconds is recorded. Example: FEVt 0.5 would be the amount 3 of air expired in one half of a second.

Arterial blood gases

What do ABGs assess?

Arterial blood gases, or ABGs, are commonly ordered for a variety of conditions. Information that includes:

- How well the patient is oxygenated (reflected in the PO_2)
- The acid-base status of the patient (reflected in the pH)
- The ability of the lungs to eliminate the cell's waste product; carbon dioxide (reflected in the Pco_2)
- The possible causes of illness (reflected in the ABG analysis)

Why arterial blood samples

Blood gases are most commonly taken on arterial blood. The reasons for this are:

- Arterial blood is maximally oxygenated; it provides information on how well the lungs are oxygenating the blood.
- Venous blood is returning to the heart; information obtained from this blood will not reflect the status of blood throughout the body. It will reflect the status of blood from one part of the body, such as an extremity.

Helpful definitions

These definitions are useful when assessing ABGs:

pH --

A value that represents the degree of acidity or alkalinity
The pH scale ranges from 0-14
Neutral pH is 7, it is neither acidic or alkaline
A pH below 7 is considered to be acidic
A pH above 7 is considered to be alkaline
The normal pH of blood is slightly alkaline, between 7.35-7.45
A blood pH below 6.8 or above 7.8 is incompatible with life

Acid --

A substance that liberates hydrogen ions (H^+) in a solution
An acid can neutralize a base
An acid has a pH 7.0

Base --

A substance that accepts hydrogen ions (H^+) in a solution
A base can neutralize an acid
A base has a pH 7.0
A base is considered to be alkaline

Pco_2 --

A value that represents the partial pressure of carbon dioxide in the blood. The normal blood value is 35-45 mm Hg
Carbon dioxide (CO_2) is eliminated by the lungs
The Pco_2 reflects the RESPIRATORY component

P_{CO_2} levels are regulated by the lungs

HCO^{3-} --

Represents the Bicarbonate content
The normal blood value is 22-26 mEq/L
HCO^{3-} level is a metabolic function of the renal system
The HCO^{3-} reflects the METABOLIC component

PO_2 --

A value that represent the partial pressure of oxygen
A PO_2 of 50 mm Hg requires supplemental oxygen

B.E. --

A value that represents the base
- sign means a deficit
+ sign means an excess

Normal arterial blood gas values

This table gives the normal values for ABGs

Measurement	**Value**
pH	7.35-7.45
P_{CO_2}	35-45 mm Hg
HCO^{3-}	22-26 mEq/L
PO_2	80-100 mm Hg
O_2 Sat	95% or
B.E.	-2 - +2

Analyzing ABGs

There are four major classifications of acid-base abnormalities:

Respiratory acidosis
Metabolic acidosis
Respiratory alkalosis
Metabolic alkalosis

Each of these types has a specific affect on the three major components of arterial blood gases.

Acid-Base Abnormality	pH	Pco_2	HCO^{3-}
Respiratory acidosis	Decreased below 7.35	Increased above 45	In normal range
Metabolic acidosis	Decreased below 7.35	In normal range	Decreased below 22
Respiratory alkalosis	Increased above 7.45	Decreased below 35	In normal h range
Metabolic alkalosis	Increased above 7.45	In normal range	Increased above 26

Consult the chart and observe:

- When the pH is decreased below 7.35 the condition is ACIDOSIS
- When the pH is increased above 7.45 the condition is ALKALOSIS
- When the Pco_2 is affected, the condition is RESPIRATORY; the Pco_2 therefore reflects the RESPIRATORY component.
- When the condition is RESPIRATORY, the HCO^{3-} is in normal range.
- When the HCO^{3-} is affected, the condition is METABOLIC; the HCO^{3-} therefore reflects the METABOLIC component.
- When the condition is METABOLIC, the Pco_2 is in normal range

The table can be further simplified by replacing the words and numbers with arrows.

Notice on this chart:

- Arrows in the same direction appear with METABOLIC conditions.
- Arrows in the opposite direction appear with RESPIRATORY conditions.

Acid-Base Abnormality	pH	Pco_2	HCO^{3-}	Direction of arrows
Respiratory acidosis	↓	↑	WNL	Different
Metabolic acidosis	↓	WNL	↓	Same
Respiratory alkalosis	↑	↓	WNL	Different
Metabolic alkalosis	↑	WNL	↑	Same

Hints for greater mastery:

- Write the normal ranges on top for the three major components, the pH, Pco_2, and the HCO^{3-}.
- Draw arrows next to the values in the problems.

Solve the following:

Place the letter that best describes the condition next to the problem.

A= Respiratory Acidosis
B= Metabolic Acidosis
C= Respiratory Alkalosis
D= Metabolic Alkalosis

	pH	Pco_2	HCO^{3-}	Answer
1.	7.30	52	24	_____
2.	7.48	45	30	_____
3.	7.47	32	23	_____
4.	7.24	63	26	_____
5.	7.57	38	35	_____
6.	7.31	40	19	_____

Answers

1. A 2. D 3. C 4. A 5. D 6. B

Did you remember the ranges for each value correctly?

pH	7.35-7.45
Pco_2	35-45
HCO^{3-}	22-26

Did you place arrows by the problems? If you did both of these, does your table look like the example?

(No arrow denotes WNL)

	7.35-7.45 pH	35-45 Pco_2	22-26 HCO^{3-}	Answer
1.	↓ 7.30	↑ 52	24	A
2.	↑ 7.48	45	↑ 30	D
3.	↑ 7.47	↓ 32	23	C
4.	↓ 7.24	↑ 63	26	A
5.	↑ 7.57	38	↑ 35	D
6.	↓ 7.31	40	↓ 19	B

Review how the problems were solved.

#1 -- is acidosis (pH is low) and the Pco_2 (respiratory component) is outside of the normal range. Thus, we have a condition that is both RESPIRATORY and ACIDOSIS or respiratory acidosis.

#2 -- is alkalosis (pH is high) and the HCO^{3-} (metabolic component) is outside of the normal range. Thus, we have a condition that is both METABOLIC and ALKALOSIS or metabolic alkalosis.

#3 -- is alkalosis with Pco_2 outside normal range, RESPIRATORY ALKALOSIS.

#4 -- is acidosis with Pco_2 altered, RESPIRATORY ACIDOSIS.

#5 -- is alkalosis with HCO^{3-} altered, METABOLIC ALKALOSIS.

#6 -- is acidosis with HCO^{3-} altered, METABOLIC ACIDOSIS.

The above problems represent simple acid base abnormalities. Analyzing the results of the arterial blood gases for these abnormalities is important. It is just as important if not more so, to understand what effect these abnormal gases have on the body, and how they became abnormal.

Causes of abnormal arterial blood gases

Causes of abnormal pH values

A. Increased pH value

In order to increase the pH value such as is found in ALKALOSIS, the body must either lose acids or increase the amount of bicarbonate in the system.

1. Loss of acids in increased pH values

a. Respiratory causes

Acid is lost when excess CO_2 is blown off by the lungs such as when the patient is breathing too fast as in hyperventilation.

Hyperventilation can occur during anxiety or extremes in emotion such as pain.

Hyperventilation can also occur when ventilator settings are delivering too many breaths per minute.

b. Metabolic causes

Acid can be lost from the GI tract during illness.

Acid is lost during vomiting (acid lost from stomach).

Acid can be lost when bile is lost (from lower GI tract).

Acid can be lost during gastric suctioning/lavage.

2. Gain in bicarbonate (HCO^{3-}) in the body in increased pH

a. Metabolic causes

Ingestion of baking soda or other alkaline substances.

B. Decreased pH value

In order to decrease the pH value (ACIDOSIS), the body must either gain acid or lose bicarbonate.

1. Lost Bicarbonate (HCO^{3-}) in decreased pH

a. Metabolic causes

Ileostomy causes body to lose HCO^{3-}

Intestinal fistulas

2. Gain in acid in decreased pH

a. Respiratory causes

Hypoventilation, patient retains acids when too little CO_2 is blown off.

Emphysema or pneumonia when gas exchange in the lungs is impaired.

Airway obstruction when no gas exchange can take place.

b. Metabolic causes

Uncontrolled diabetes with fat breakdown and metabolism (ketoacidosis)

Starvation diets (breakdown/metabolism of fatty acids)

Poor renal function (increase in BUN and creatinine)

Cardiac disorders (poor circulation/hypoxia)

Overdose of salicylates (aspirin)

The respiratory and renal system are the main mechanisms responsible for maintaining our bodies delicate acid-base balance between 7.35 and 7.45 pH.

This acid-base balance is maintained by keeping the ratio of HCO^{3-} and CO_2 constant within the body. The normal ratio is 20 HCO^{3-} molecules for every 1 CO_2 molecule. Written as:

$$20\ HCO^{3}- : 1\ CO_2.$$

Any change in the body's pH changes this ratio, and the body must make an effort to compensate or correct the imbalance.

The role of the lungs in compensation for acid-base abnormalities

The lungs role in maintaining this balance includes:

- The ability to retain or blow off CO_2 by changing the respiratory rate and depth.
- Retaining CO_2 by breathing less frequently (hypoventilation) would decrease the body's pH.
- Blowing off CO_2 by breathing more rapidly (hyperventilation) would increase the body's pH.
- The lungs are able to compensate for pH changes faster than the kidneys.

The kidney's role in compensation for acid-base abnormalities

The kidney's role in maintaining this balance includes:

- The ability to retain or excrete bicarbonate (HCO^{3-}).
- Retaining bicarbonate would increase the body's pH.
- Excreting bicarbonate would decrease the body's pH.

The table below divides into sections, one for each of the major acid base abnormalities. Each section gives the cause, signs and symptoms and method of compensation the body will use to correct the imbalance.

HOW BODY COMPENSATES FOR ACID-BASE IMBALANCES

Respiratory acidosis

Cause	Signs & Symptoms	Compensation
Hypoventilation Pneumonia Emphysema Asthma Drug overdose Head injury Reye's syndrome CHF	Decreased respiratory rate Weakness Disorientation Drowsiness Headache Blurred vision Coma	Lungs Hyperventilation Kidneys retain bicarbonate and secrete hydrogen ions

Metabolic acidosis

Cause	Signs and Symptoms	Compensation
Diabetes Starvation diets Poor renal function Systemic infections Tissue anoxia	Weakness Confusion Stupor Kussmaul breathing Unconsciousness Cardiac arrhythmias Hypokalemia Increased pulse Decreased B/P	Lungs Hyperventilation Kidneys retain bicarbonate Secrete hydrogen ions Retain potassium Retain sodium

Respiratory alkalosis

Cause	Signs & Symptoms	Compensation
Hyperventilation Anxiety Pulmonary emboli Fever Salicylate overdose	Decreased respiratory rate Hypoventilation (late) Numbness/tingling in extremities Convulsions (late) Tetany	Lungs Hypoventilation Kidneys excrete bicarbonate Retain hydrogen ions

Metabolic alkalosis

Cause	Signs & Symptoms	Compensation
Vomiting Ingestion of alkaline substances Cushing's disease Corticosteroids Gastric suctioning	Hypoventilation Confusion Irritability Tetany (late)	Lungs Hypoventilation Kidneys excrete potassium sodium bicarbonate Retain hydrogen

Arterial blood gas values in compensation

When the body is compensating for an acid-base imbalance it may be difficult to discover the original problem. The reason detection is sometimes difficult is that not all of the values are always abnormal.

- The pH may be near normal or within the normal range.
  pH 7.36 Pco_2 50 HCO^{3-} 29

When a normal pH is noted along with an abnormal Pco_2 or HCO^{3-}, the body is compensating for an acid-base imbalance. When this occurs follow these simple steps to detect the original problem:

- First check which end of the scale the pH is nearest, acidosis or alkalosis. This will reveal the basic problem.
  7.36 is low normal and therefore acidosis was the primary problem.
- Next check for increase or decrease in the other values.
  Pco_2 was increased to 50, HCO^{3-} was increased to 29.
- Check which type of acidosis has an increased value in Pco_2 or HCO^{3-}.
  Out of two types of acidosis, only one has an increased Pco_2, respiratory acidosis.
  Metabolic acidosis does not have an increased HCO^{3-}, therefore the body must be trying to compensate for the respiratory acidosis by retaining more bicarbonate.

Go through another mixed acid-base problem:

- First check which end of the scale the pH is nearest, acidosis or alkalosis.
  pH 7.36 Pco_2 26 HCO^{3-} 12
  For this example the basic problem was acidosis.
- Next check for increase or decrease in other values.
  Pco_2 decreased to 26, HCO^{3-} decreased to 12.
  Out of two types of acidosis, neither one has a decreased Pco_2.
  Metabolic acidosis has a decreased HCO^{3-}.
  The original disorder was metabolic acidosis.
  The body must be trying to compensate for the metabolic acidosis by increasing the rate of respirations which would blow off CO_2 resulting in a decreased Pco_2.

Acid-base imbalance problems for practice

Try the following problems, use a letter to answer the questions:

A= Respiratory Acidosis B= Metabolic Acidosis
C= Respiratory Alkalosis D= Metabolic Alkalosis

Which condition is most likely to develop with:

#1. Continuous nasogastric suctioning ____
#2. Anxiety ____
#3. Airway obstruction ____
#4. Diabetic acidosis ____
#5. Salicylate toxicity ____
#6. Severe vomiting ____

The following laboratory values show which condition:

	pH	P_{CO_2}	HCO^{3-}	Answer
#7.	7.20	65	24	_____
#8.	7.20	45	18	_____
#9.	7.60	25	23	_____
#10.	7.33	49	24	_____
#11.	7.30	40	15	_____
#12.	7.36	52	30	_____
#13.	7.43	49	32	_____

Answers to problems

1. D
2. C
3. A
4. B
5. C
6. D
7. A
8. B
9. C
10. A
11. B
12. A, compensated
13. D, compensated

Interventions for acid-base disorders

Once the type of acid-base disorder has been determined, these interventions should be considered:

Interventions for acid-base disorders

Disorder	Intervention	Rationale
Respiratory Acidosis	Bronchodilators	To open the airway
	Oxygen	Increase O_2 concentration
	Frequent vital signs	Respiratory depression may occur
	Turn cough, and deep breathe	To promote lung expansion
	Force fluids	To thin secretions
	Assess LOC	Decreased LOC may occur
	Administer narcotics, sedatives & hypnotics with caution	To prevent CNS depression
Metabolic Acidosis	Assess for dehydration	Assess for hypovolemia
	Strict I & 0; V/S	To protect patient
	Seizure precautions	To assess for change in
	Neurological checks	neurological status
		Hypokalemia may occur
	Kayexalate for hypokalemia	Cardiac arrhythmias may occur
	Monitor ECG	
Respiratory Alkalosis	Rebreathing mask if due to hyperventilation	CO_2 will be inhaled paper bag will work if mask is not available
	Assess for cause of hyperventilation	Pain, anxiety, head injury may cause hyperventilation,
	Administer pain meds prior to intense pain	To reduce pain and anxiety, promote relaxation,
	Provide back rubs	
	Assess for hypokalemia	Due to cellular buffers
	Assess for headache paraesthesia, tetany	Due to vasoconstriction
Metabolic Alkalosis	Assess vital signs	Decreased respirations may occur
	Administer Diamox as ordered	To help kidney excrete HCO^{3-}
	Administer Chloride or	To help kidney excrete HCO^{3-}
	K+ and Cl- or KCl if ordered	Serum K+ may be decreased
	Assess for tetany	Metabolic Alkalosis (late)

Section IV -- PROCEDURES USED IN THE TREATMENT OF RESPIRATORY PROBLEMS

Oxygen administration

Oxygen is an essential element. The air we breathe is about 21 percent oxygen. The air exhaled is about 16 percent oxygen. Often, when a patient is receiving supplemental oxygen, you will record the percentage of oxygen being delivered as 30 percent, 40 percent. If the patient is on 21 percent oxygen they are receiving the oxygen concentration of room air.

Oxygen is often administered as a drug to treat oxygen deficiency, or hypoxia. Hypoxemia is a deficiency of oxygen in the blood. The indications for oxygen use are related to hypoxia or hypoxemia.

Indications for oxygen use

The single most reliable indicator of the need of oxygen is the PaO_2. The PaO_2 is the partial pressure of arterial oxygen. When this is decreased, hypoxemia is present. This may occur long before the signs and symptoms of hypoxemia are present.

If the patient has arterial blood gases drawn and the PaO_2 is below 50 mm Hg, that patient has hypoxemia and needs oxygen. The normal PaO_2 range is 80-100 mm Hg.

Even before the results of the arterial blood gas testing are back the patient may exhibit signs and symptoms of hypoxemia.

Signs and symptoms of hypoxemia are:

Dyspnea (Shortness of breath, difficulty breathing)
Cyanosis (bluish tinge to lips, mouth, or nailbeds)
Anxiety, or feelings of impending doom
Restlessness
Confusion
Tachycardia (rapid heart beat) with elevated blood pressure
Pale, cool extremities (due to vasoconstriction)

If your patient exhibits these signs and symptoms notify the physician at once, get a blood gas reading and administer oxygen as ordered.

Conditions where supplemental oxygen may be required:

- Anemia (severe)
- Carbon monoxide poisoning, oxygen always required initially
- Congestive heart failure (CHF)
- Chronic obstructive pulmonary disease (COPD)
- Myocardial infarction

- Shock
- Pulmonary edema
- Pneumonia
- Surgical procedures

Prior to administration of oxygen:

- Assess the patient's respiratory condition:
  Respirations or respiratory rate
  Difficulty breathing
  Cyanosis of mucous membranes or nailbeds
  Results of arterial blood gases if available
- Obtain and record the baseline vital signs
- Check physician's order for method of oxygen administration
  Face mask
  Nasal cannula
  Oxygen tent
  Ventilator
- Check physician's order for rate of oxygen delivery
- Call respiratory therapy or gather supplies
  Wall oxygen unit or portable oxygen bottle
  Mask, cannula or other delivery method
  Distilled water and humidifier unit for oxygen set-up
- Remove fire hazards from area
- Remove unnecessary electrical equipment
- Post "NO SMOKING, OXYGEN IN USE" sign as needed. Most hospitals are now "no smoking" in all areas

METHODS OF OXYGEN ADMINISTRATION

Method	Comments
Cannula	Nasal cannula consists of a length of tubing with two flexible curved nasal prongs that fit into nares. Nasal prongs should curve with the nasal passage as oxygen is delivered via nasal prongs. Delivers 24% oxygen at 2 L/minute. Flow rates greater than 5 L will dry the nasal membranes. Patient should breathe through the nose to prevent a loss of oxygen. Lubricating the nares will help prevent irritation from cannula
Mask	Face mask needs to cover both the mouth and nose. Mask should fit tightly enough to prevent dilution of oxygen with room air. A flow rate of 8-15 L/minute delivers 40-60% oxygen concentration.

No matter which method of oxygen administration is used, oxygen must always be humidified prior to use to prevent drying of the mucous membranes. Check agency policy for type of solution to be used. Distilled water or normal saline are commonly used.

Oxygen administration and the patient with chronic carbon dioxide retention:

Patients who have chronic pulmonary disease such as emphysema have become conditioned to high levels of carbon dioxide. Their respiratory regulatory center in the brain is no longer using the level of blood carbon dioxide as an indicator of a need for oxygen. For these persons the respiratory regulatory center is dependent upon this hypoxemia. If the oxygen level is suddenly increased, the stimulus to breathe is depressed. In other words, high levels of oxygen will decrease respirations. Persons with chronic carbon dioxide retention need a much lower concentration of oxygen and must be carefully monitored.

Complications of oxygen administration

Oxygen, like any other drug, can become toxic if too much is administered. These are signs and symptoms of oxygen toxicity:

- Nausea and vomiting
- Restlessness
- Pallor
- Paresthesia
- Sore Throat
- Substernal chest pain

Oxygen toxicity that continues unchecked can lead to permanent brain and lung damage. Premature neonates exposed to high concentrations of oxygen may develop retrolental fibroplasia, a serious eye defect, that can cause visual impairment or blindness.

Chest tubes

Chest tubes are used to remove fluid, blood and/or air from the pleural cavity. The pleural cavity normally contains no air or blood and is located between two layers of tissue, the visceral pleura, and the parietal pleura. The pleural cavity has a thin layer of lubricant that allows the pleura to slide and move with each other. The pleural cavity works like a vacuum, as it holds the pleura together during and between respirations.

During inspiration, when the diaphragm contracts and moves down, creating more lung space, the pleura expand together.

During expiration when the diaphragm relaxes, the pleura prevent the lung tissue, which is very elastic, from collapsing. Any separation of the two pleura destroys the vacuum and prevents the two layers from moving as a unit. When the vacuum is destroyed, the lungs will collapse and the air available for gas exchange will decrease.

Indications for chest tubes

Closed pneumothorax:

In a closed pneumothorax there is no air exchange between the external environment and the lungs. When the visceral pleura, which is the pleura closest to the lungs, is damaged, air and or fluid/blood can enter the pleural cavity, but it cannot exit. The vacuum is destroyed, and the lung collapses.

During inspiration, air moves into the pleural cavity; during expiration, not all of the air moves out. During the next respiratory cycle the lung cannot expand fully due to this trapped air. This process is repeated with the area in the pleural cavity growing until a large portion or all of the lung is collapsed. A closed pneumothorax can result from:

- Rupture of an abscess or bleb on the lung's surface
- Puncture by fractured rib
- Ruptured bronchus or perforated esophagus
- External trauma followed by airtight seal to chest
  - Penetration during thoracentesis
  - Penetration during central vein catheter placement
  - Pulmonary barotrauma from mechanical ventilation

Open pneumothorax:

An open pneumothorax occurs when the chest wall, parietal pleura, and visceral pleura are entered from an external source allowing air to enter the lungs. This entry into the lungs destroys the vacuum. An open pneumothorax can result from chest trauma such as a knife wound, gunshot, or an auto accident. An open pneumothorax can also occur during surgery after removal of a lung or when a closed pneumothorax is being corrected with chest tube placement.

Hemopneumothorax

Another term often used with pneumothorax is hemopneumothorax, which means there is more blood trapped in the pleural cavity than air.

How chest tubes work

The purpose of chest tubes is to restore the normal lung pressure, and/or to drain air or fluid from the pleural cavity. When this is accomplished, the lung can reexpand. One end of the chest tube is inserted into the chest wall, the other is connected to a chest drainage system. The most commonly used is the disposable unit system.

This chest drainage system has three compartments:

1st Compartment -- Collects any fluid leaving the lung and is known as the "Collection Chamber."

2nd Compartment -- Prevents any air from the atmosphere from entering the lung through the chest tube. This chamber has water in it; the water allows air from the lung via the tube to pass by and prevents air from the atmosphere from entering. To understand this concept, think of a straw. A straw allows you to blow air through it but the end of the straw is underwater and no air can enter this end until all of the water is gone. This compartment is known as the "Water Seal Chamber."

3rd Compartment -- This compartment also contains water. However, the water determines the degree of suction. The suction is needed to remove the air and/or fluid from the pleural cavity. The amount of suction used can be changed by changing the amount of water in this compartment. Although this bottle connects to a suction control unit, the amount of suction depends on the level of water in this compartment. This compartment is known as the "Suction Control Chamber".

The chest tube system is simple in its principle. It removes unwanted substances from the pleural cavity and prevents their reentry so that the lung can re-expand.

Assisting with chest tube insertion

Prior to the procedure

- Assess reason for chest tube insertion
- Assess and record respiratory condition to include respiratory rate, rhythm and breath sounds
- Decreased or absent breath sounds may be present when air or fluid is trapped in the pleural cavity
- Check physician's orders for time of insertion, premedication, and any special instructions
- Gather equipment

  Sterile chest tube insertion tray with:

  Antiseptic (povidone-iodine)
  Sterile chest tube
  Chest drainage system unit
  Sterile petrolatum gauze
  Surgical blade and other tools
  Sterile water or normal saline to keep at bedside
  Premedication (sedative and analgesic may be ordered)
  Local anesthetic (lidocaine)
  Covered hemostat to keep at bedside
- Check for signed informed consent
- Administer any premedications ordered at correct time
- Set up the chest drainage system as ordered, maintain sterility of system, follow the directions on the system

During insertion

- Patient should be in the high-Fowler's position
- Incision will be made by physician to place chest tube

  Midclavicular line high on chest if only air is to be removed
  Midaxillary line low on chest if blood, fluid alone, or blood and fluid with air is to be drained
- Patient should be instructed not to cough
- Instruct patient pain is temporary as the pleura are difficult to anesthetize
- Once the tube has been inserted into the chest wall, then it is connected to the drainage system using a barrel connector
- The chest tube is sutured to the chest wall, and the petrolatum gauze is placed around insertion site
- The entire procedure is performed under sterile conditions to prevent infections in the pleural cavity

Post procedure

- Assess respirations for rate and rhythm. Assess breath sounds for any change
- Assess for gentle bubbling in the suction control chamber
- Assess for gentle bubbling in the water-seal chamber; this should be present if the purpose is to remove air
- Assess drainage collection chamber for type and amount of drainage. If drainage is bloody, assess amount and report
- Tape connections tightly with tape to prevent disconnection
- Assess dressing for drainage and airtight seal

Respiratory distress may indicate an increased accumulation of air or fluid in the pleural cavity. A tension pneumothorax may have occurred; this is the most serious complication of chest tubes.

A tension pneumothorax occurs when air is trapped in the lungs and cannot escape. With each breath, the space is under more pressure. Eventually, the heart and great vessels are pushed to the opposite side by the increasing pressure. This is a life-threatening situation because the ability of heart to receive and pump blood is compromised. Decreased cardiac output with a drop in blood pressure will occur if not treated.

The signs and symptoms of tension pneumothorax are:

- No breath sounds on affected side
- The affected side will appear inflated with air, but does not fall on expiration
- The trachea is displaced to the unaffected side
- The pulse will be thready and fast
- Serve dyspnea with sudden sharp chest pain
- Diminished to absent motion and breath sounds of affected side
- The affected side will appear inflated upon expiration
- Tympany on percussion of chest; diminished or absent tactile fremitus
- Heart beat displaced secondary to mediastinum shift
- Trachea is displaced to the unaffected side
- Onset of shock with falling pressure and thready pulse
- X-ray of chest confirms pneumothorax

Respiratory isolation

Respiratory isolation is used to stop the transmission of pathogens from the respiratory tract of an infected person to an uninfected person. Transmission of pathogens can occur by direct contact or by the airborne route and is primarily used in childhood diseases.

Direct contact	Airborne route
Soiled linen	Droplets coughed
Soiled tissue	Droplets sneezed
	Droplets breathed

Examples of diseases where respiratory isolation would be required include:

- Herpes Zoster
- Measles (both varieties)
- Meningococcal infections
- Mumps
- Pertussis

The following measures are taken in respiratory isolation to prevent the transmission of the disease:

- Private room
- Door is kept closed at all times
- Masks should be worn when entering room, especially if patient is coughing
- Gloves/gowns are needed if contact with infected sputum or body secretions is likely
- Contaminated materials such as tissues or linen are placed in plastic bag, sealed and placed in another bag for contaminated materials as per agency policy prior to removal from room.

Section V -- DIETS

There are no diets specific for this system. High calorie diets because so much energy is required for breathing may be recommended. Fluids, up to 3000 ml per day, may be recommended.

Section VI -- DRUGS

The following tables give general information; consult a drug handbook prior to administering any unfamiliar drug.

ANTIHISTAMINES

Action: To block the effects of histamine, thereby, reducing the edema caused by the release of histamine.

Indications: To treat allergies, colds, rhinitis, allergic reactions.

General comments: Drowsiness is a common side effect, and the drug may be prescribed for sedation at bedtime; caution against driving or other activities were mental alertness is required. Alcohol and other sedatives should be avoided while taking this medication. Dry mouth and thickened bronchial secretions may occur. Development of tolerance is common over long term use.

Examples of drugs in this classification:

Generic	Trade	Comments
Brompheniramine	Dimetane Bromamine Brombay Bromphen	Given PO, SC, IM, IV May be taken with food if stomach upset occurs Administer IV, slowly Notify physician if high fever, chills or mouth ulcers occur
Dimenhydrinate	Dramamine Dramilin Hydrate Nauseatol Motion-Aid	Given PO, IM, IV Used often for motion sickness or as an antiemetic Given prior to radiation therapy Effects last 3-6 hours May be purchased over-the-counter
Diphenhydrinate	Benadryl Allerdryl Benylin Benojet Nordryl Insomnal	Given PO, IM, IV Used often for anaphylaxis Given as pre-medication prior to chemotherapy or blood Purchased over-the-counter Used for sedative effect
Hydroxyzine Hydrochloride	Vistaril Atarax	Given PO, IM Tablets may be crushed Preferred IM site is deep such as upper outer quadrant of buttocks

Promethazine Hydrochloride	Phenergan Promine Prothazine Pentazine	Given PO, IM, IV Tablets may be crushed Given as pre-medication prior to chemotherapy or blood IM injection as above

ANTITUSSIVES

Action: To suppress coughing, may be narcotic or non-narcotic.

Indications: Non-productive cough or overactive cough.

General comments: Narcotic antitussives are addictive, may cause constipation and drowsiness. Alcohol should be avoided.

Examples of drugs in this classification:

Generic	Trade	Comments
Codeine	Codeine	Given PO, SC, IM, Narcotic antitussive Administer with food or milk to decrease GI distress, Nausea is a common side effect Addiction and dependence may occur
Dextromethorphan Hydrobromide	Hold Sucrets D.M. syrup Formula 44 Pertussin 8 hour Congespirn Contact severe cold formula Cold histamines formula Hycodan Novahistine cold formula NyQuil Triaminic-DM	Non narcotic antitussive Given PO Purchased over-the-counter Available in combinations with decongestants, analgesics
Diphenhydramine Hydrochloride	See previos page under Antihistamines	

BRONCHODILATORS

Action: To relax bronchial muscle, reduce bronchial edema and mucous production.

Indications: To control wheezing, and improve activity tolerance by reducing respiratory symptoms. May be used in allergies, asthma, bronchitis, emphysema, hay fever, or obstructive airway disease.

General comments: Bronchodilators increase cardiac rate and output palpitation, nervousness, restlessness, anxiety, and hypertension are side effects. Vital signs should be taken prior to and frequently during administration by parental routes. Bronchodilators given by the inhalation route are often misused, and patient should be instructed in the correct procedure, dosage, frequency of use.

Examples of drugs in this classification

Generic	Trade	Comments
Ephedrine	Ephedrine Gluco-Fedrin	Given SC, IM, IV slow Intranasal, and topical Common ingredient in over-the-counter cold/allergy medications. Check for drug interactions when multiple medications are ordered
Epinephrine Hydrochloride	Adrenaline Epifrin	Given by SC, inhalation, topical routes, IM, IV Always check for correct route on Bronkaid Mist orders and medication vial Primatene Mist Always aspirate prior to injecting to prevent accidental IV administration of SC dose, Avoid buttocks for IM route, Cardiac monitor is recommended for IV The first line drug in cardiac arrest
Isoproterenol Hydrochloride	Isuprel Aerolone Norisodrine	Given inhalation, SL, rectal, SC, IV, IC Sublingual tablet given by rectal route if so ordered Given via oxygen aerosol treatments Do not administer with Epinephrine, give at least 4 hours apart Given for cardiac arrest

Terbutaline Sulfate	Brethine Brethaire Bricanyl	Given by PO, SC, inhalation, IV routes Tablets may be crushed Tablets can be taken with food for GI upset
Theophylline	SloBid Theo-Dur Bronkodyl Lanophyllin	Given PO route Therapeutic range is 10-20 mcg/ml Enteric coated tablets should not be crushed (or sustained release tablets) Administer after meals with water to prevent GI irritation Dizziness occurs frequently in early treatment

DECONGESTANTS

Action: To relieve nasal congestion.

Indications: Allergies, upper respiratory infections, otitis media.

General comments: May have stimulant effects, do not administer at bedtime. Often found in over-the-counter medications especially antihistamines. Avoid mixing OTC drugs as action may be increased.

Examples of drugs in this classification

Generic	Trade	Comments
Ephedrine	Efedron Nasal Vatronol	See bronchodilators May produce sneezing, burning of nasal passages
Oxymetazoline	Afrin Dristan Duration Neo-Synephrine Sinex	Available over-the-counter Prolonged decongestant effect Rebound congestion can occur with misuse Given by nasal drops
Pseudoephedrine	Neofed Novafed Sudafed Sudrin	Given PO Tablets may be crushed See also Ephedrine

EXPECTORANTS

Action: To decrease the viscosity of respiratory secretions.

Indications: Non-productive coughs.

General comments: Use and therapeutic value are controversial. Any persistent cough should be evaluated for cause, as it may indicate a more severe problem. Increased fluid intake and humidification of air is recommended to augment effects.

Examples of drugs in this classification

Generic	Trade	Comments
Guaifenesin	Anti-Tuss Baytussin Robitussin	Given PO Common ingredient of cough medications Causes nausea and vomiting occasionally
Iodinated glycerol	Organidin	Given PO Increase fluid intake
Terpin hydrate		Given PO Increase fluid intake

SECTION IV -- GLOSSARY

Alveoli	Small air sacs within the lungs where internal respiration (gas exchange) occurs
Alveolus	One air sac (singular)
Apnea	Break or period of no breathing
Atelectasis	Collapse of lung tissue
Bronchiectasis	Chronic abnormal state with spasms of coughing, purulent sputum due to chronic dilation of the bronchus
Bronchoplasty	Repair of the bronchus
Bronchospasm	A spasm of the muscles located around the bronchi
CO_2	Carbon dioxide, the odorless, colorless gas which is a cellular waste product expelled by the lungs
COPD	Chronic Obstructive Pulmonary Disease, term for obstructive pulmonary disorders such as emphysema and bronchitis
Diaphragm	Muscle that separates the thoracic and abdominal cavities. The primary muscle used in respiration. When the diaphragm contracts, the lungs expand and inspiration occurs; when it relaxes, expiration occurs.
Dyspnea	Difficult breathing
Emphysema	Obstructive lung disease where the alveoli increase in size due to damage or destruction
Epiglottis	The flap-like structure covering the larynx during swallowing
Epistaxis	Bleeding from the nose, nosebleed
Eupnea	Normal breathing, no distress in breathing

Hemopneumothorax	Blood in the pleural cavity
Hemoptysis	Blood in the sputum
Hyperventilation	Rapid breathing which is prolonged and deep
Hypoventilation	Slow, shallow breathing
Larynx	Also called the voice box. Located at the upper end of the trachea, the larynx contains the false and true vocal cords which aid in speech
Laryngitis	Inflammation of the larynx
Nares	Nostrils
O_2	Oxygen, an essential element for life
Pertussis	Whooping cough
Pharynx	An upper airway tube that extends from the nasal cavity (and oral cavity) to the larynx, divided into three parts: Nasopharynx-upper portion above the palate Oropharynx-palate to hyoid bone Laryngopharynx-hyoid bone to Larynx
Pleural cavity	The space (potential not actual) between the two pleural layers of the lung, the parietal pleura and the visceral pleura
Pleurisy	Inflammation of the pleura
Pneumonia	Congestion and inflammation of the lungs
Pneumohemothorax	Air and blood in the pleural cavity
Pneumothorax	Air in the pleural cavity
Pulmonary edema	Fluid in the lung tissues
Respiration	The act of breathing, the exchange of gases between the atmosphere and lungs

Rhinitis	Inflammation of the nasal membranes
Rhinoplasty	Cosmetic procedure for repair of nose, also called a "nose job"
Thoracentesis	Drainage of fluid from the chest wall via a puncture wound (needle)
Trachea	An airway tube extending from the larynx to the right and left bronchi, also called the windpipe

Neurologic System

NEUROLOGIC SYSTEM
Table of contents

Section I -- OVERVIEW

Primary functions

1. Controls and manages most body functions needed for survival.

Receives, decodes, interprets messages from the body's internal environment to include (list is not all-inclusive):

Changes in:
Respiration and heart rate
Levels of circulating hormones and enzymes
Core temperature
Acid-base balance of the blood

Transmits messages to the cells of the body including:

Initiation and continuation of respiration via the stimulation of the intercostal muscles and diaphragm.
Adjustments in amount and types of secretions from glands.
Transmits electrical impulses for control of muscular movement, coordination, balance and posture.

2. Allows for interaction and communication with the external environment by receiving, decoding, and interpreting messages such as:

Changes in external environment to include temperature, level of safety and sensations of movement.
Sensory stimulation to include olfactory, auditory, tactile, gustatory or visual stimuli.

3. Allows for the creation, development, perception, transmission and evaluation of thoughts and ideas.

Ability to send and receive messages (communication) via speech, gesture, written word, or technological advancement (telephone, telegraph, computer) to another human being.
Ability to learn concepts and use these concepts to create ideas or products.

4. Responsible for the individual distinct characteristics such as temperament, feelings, cognition, and personality that make each person unique.

Components and function

This system contains the central nervous system (CNS), peripheral nervous system (PNS), and the sensory organs. The CNS contains the brain and the spinal cord. The PNS contains the cranial nerves, spinal nerves and the autonomic nervous system which contains both the sympathetic and parasympathetic nervous system. The sensory organs included are the eye and ear.

Central nervous system (CNS)

The CNS is a continuous system that is divided into two major components: the brain and spinal cord. The CNS has a protective covering composed of 3 membranes called meninges: the dura mater, pia mater and arachnoid membrane. Cerebral spinal fluid is produced by the choroid plexus (capillaries and ependymal cells) ventricles and circulated around the brain's surface and in the spinal cord.

Brain

The brain is a complex organ with a multitude of components, this reference will discuss four major parts of the brain (Fig. 3A):

I. Brain stem
II. Cerebellum
III. Diencephalon
IV. Cerebrum

Major Parts of the Brain

Name	Function/Comment
I. Brain stem	Interconnects the spinal cord, cerebellum, diencephalon, and cerebral cortex. Controls basic body functions needed for survival (respiration, cardiovascular activity, and movement of eyeballs, head and trunk movement) Individual may live with all parts of the brain destroyed except the brain stem Ten of the twelve pairs of cranial nerves originate here. Has three components: medulla oblongata, pons, and the mesencephalon.
Medulla oblongata	A continuation of the spinal cord into the skull forming the lower brain stem. Cardiac center: Regulates the rate and contraction force of the heart. Also regulates diameter of blood vessels. Respiratory center: Regulates rhythm. Control center for basic life-support reflexes: Coordinates swallowing, gagging, coughing, sneezing, hiccupping, vomiting, and salivating.

The Brain and Meninges

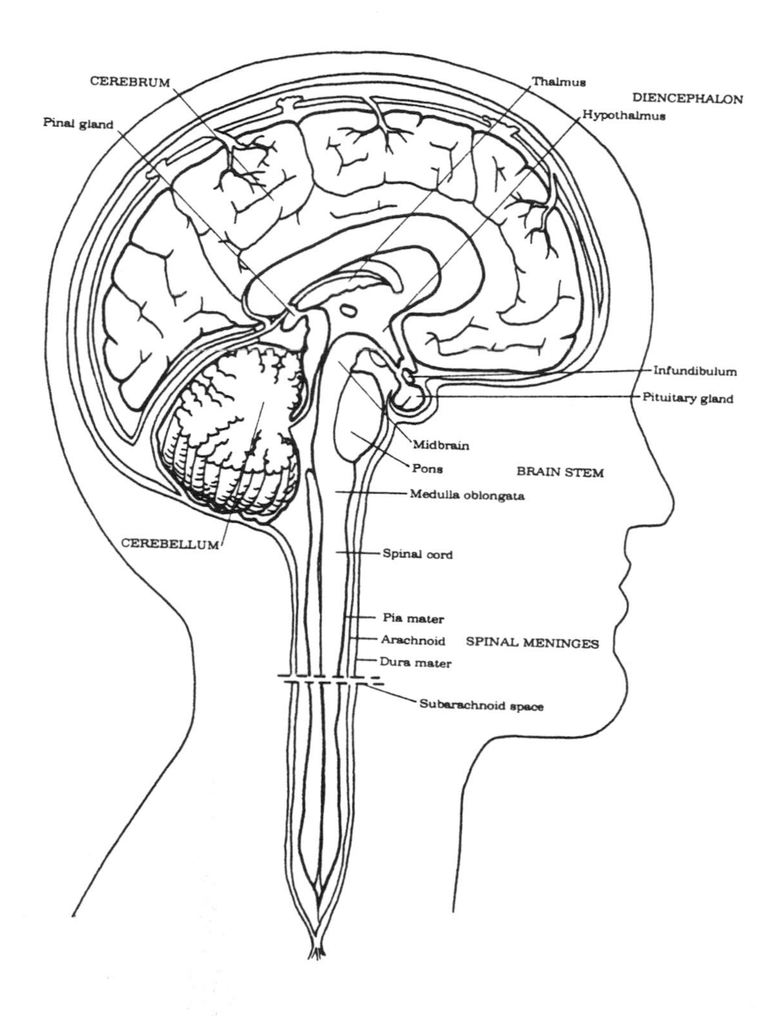

Figure 3A

The pons	Function primarily as a bridge between cerebellum the cerebrum and the brain stem. It is the connection between the upper and lower levels of the CNS.
Mesencephalon (midbrain)	Regulates pupillary size and lens shape. Involved in the function of localizing sound sources.
II. The cerebellum	Coordinates motor movement with visual, auditory, somesthetic and vestibular information to produce smoothly controlled movements in skeletal muscle. Needed for coordination, balance, posture and normal muscle tone. New studies suggest that rote movements such as touch-typing or finger movements to play an instrument also originate in the cerebellum.
III. The diencephalon	Composed of the dorsal thalamus, hypothalamus and subthalamus.
Thalamus	Commonly referred to as the thalamus. Sensory stimuli (except olfactory) are received here and then relayed to cortical areas. Thalamus makes the sensation of pain, light touch and pressure and temperature possible.
Hypothalamus	Homeostatic control center: integrates information from external and internal environment and then stimulates other brain parts to respond accordingly. Fluid balance, body temperature, hunger, and the major aspects of hormonal control are located in the hypothalamous itself.
V. Cerebrum	Forms the majority of brain tissue. Contains a right and left cerebral hemisphere interconnected by the corpus callosum. Each cerebral hemisphere contains white matter, gray matter, and basal ganglia. Each hemisphere contains four different lobes; frontal, temporal, parietal and occipital named after bones of the skull.
Frontal lobe	Site of personality and intellectual functioning. Area where ideas are developed, plans are made for the future. This area also controls movement and produces speech.
Temporal lobe	Auditory cortex; allows for discrimination of sound frequency. Needed to determine the meaningfulness of sounds and speech. Long-term memory recall is stored here.
Parietal lobe	Primary sensory cortex; allows for discrimination of sensory input such as location, and intensity. Is needed to feel vibration, to determine the position or movement of body parts.
Occipital lobe	Primary visual cortex; interprets visual input.

Spinal Cord

The spinal cord is the link between the brain and the peripheral nerves, and extends from the foramen magnum at the base of the skull to the conus medullaris located at the first or second lumbar vertebra. The cord itself is made up of both white and gray matter and is housed in the bony vertebral canal. The white matter contains both sensory and motor tracts.

Peripheral nervous system (PNS)

All nerve impulses are transmitted to and from the CNS by the PNS which is composed of the cranial nerves, spinal nerves and the autonomic nervous system.

Cranial nerves (CN)

There are 12 cranial nerves that eminate from the base of the brain. See Table 3-1 for a description of the cranial nerves.

CN	Name	Function	Malfunction Indications
I	**Olfactory**	Perception of smell	Loss or disturbance in perception of smell
II	**Optic**	Vision	Blindness may be partial or complete depending on lesion location
III	**Oculomotor**	Movement of both eyes to center, upper right and upper left position Movement of left eye to side right and to lower left position Movement of right eye to side left and to lower right position Regulates pupillary size Raises eyelid	Deviation of eye outward Double vision Dilatation of pupil Ptosis (drooping eyelid)
IV	**Trochlear**	Movement of left eye to lower right and left	Deviation of eye upward and outward
V	**Trigeminal**	Sensory input to face Chewing	Pain or sensation loss in face, forehead, eye or temple Difficulty chewing

VI	**Abducens**	Movement of right eye to side right Movement of left eye to side left	Deviation of eye outward Double vision
VII	**Facial**	Facial expression Glandular secretion Taste	Paralysis of one side of face Loss of taste in foods
VIII	**Acoustic** Vestibulocochlear	Hearing Sense of equilibrium	Ringing in ears Deafnessness, dizziness Nausea or vomiting
IX	**Glossopharyngeal**	Taste perception Ability to swallow	Altered taste perception Difficulty in swallowing
X	**Vagus**	Involuntary muscle and gland control Swallow and speech Taste	Hoarseness Difficulty in swallowing Difficulty with speech
XI	**Spinal accessory**	Movement of shoulder and head Ability to swallow Speech	Drooping of shoulder Loss of full ROM in head
XII	**Hypoglossal**	Tongue movement	Deviation of tongue Thick speech Paralysis of one side of tongue

A memory aid for learning the cranial nerves is:

On Old Olympus Towering Tops A Finn And German Viewed Some Hops.

Each letter corresponds to the first letter of the nerves in order from I-XII.

Spinal nerves

The spinal nerves branch off the spinal cord. The spinal cord has 31 pairs of spinal nerves; 8 pairs of cervical nerves, 12 pairs of thoracic nerves, 5 pairs of lumbar nerves, 5 pairs of sacral nerves and 1 pair of coccygeal nerves. These nerves provide a pathway for feedback to and from the brain. They alert the brain of sensory changes such as hot/cold temperatures, pressure from touch, and pain. Figure 3B shows how the spinal nerves innervate various parts of the body: the

Nerves of the Body

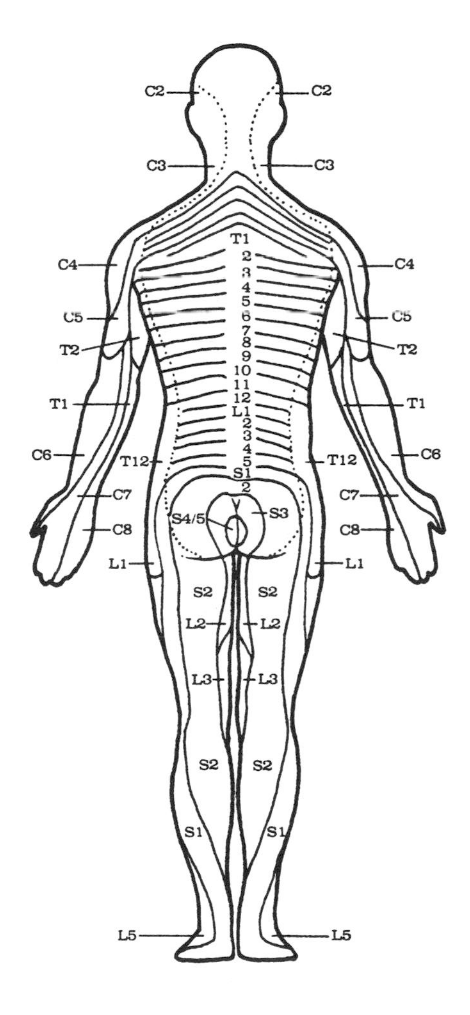

Figure 3B

numbers on the regions correspond to the spinal nerve (example: T stands for thoracic, the number indicates which thoracic level nerve).

Sensory information will be transmitted by a nerve from the particular peripheral area to the spinal cord and finally to the brain where the nerve impulse is decoded and interpreted. Information is transmitted via chemical messengers called neurotransmitters. The ability to detect pain, cold, heat, and pressure depend on specific sensory neurons used for that specific task. Referred pain occurs when areas of injury share the same innervation pathway. An example of referred pain is the shoulder and arm pain some patients feel during a heart attack. Phantom pain (pain felt in an amputated limb) may be due to the stimulation of the remaining nerve pathways in the stump. The neuron stimulation is transmitted and the cerebral cortex identifies the input as "pain" from memory and the patient complains of feeling pain in a limb that is no longer present.

Autonomic nervous system

The autonomic nervous system (ANS) is in control of involuntary bodily functions such as the regulation of the smooth muscle, cardiac muscle and gland function (salivary, gastric, sweat). The ANS is divided into the sympathetic and parasympathetic systems. The **sympathetic nervous system** enables the body to respond to stress by increasing the heart rate and blood pressure to prepare for a "flight or fight" response. Blood is deverted from organs that are non-vital in an emergency such as the skin, bladder, and stomach to organs that survival may depend on such as skeletal muscle, heart and brain. The **parasympathetic nervous system** is a restorative system that can increase glandular activity (such as saliva production) to facilitate organ function (such as digestion).

SENSORY ORGANS

Ear

The ear is the vehicle for receiving auditory stimulation. The ear consists of three principal parts (Fig 3C): The outer, middle and inner ear.

The outer ear has two parts, the pinna and the ear canal. It is responsible for the following functions:

Protection of the ear drum
Maintenance of a constant temperature
Collection of sound waves from the environment

The Ear

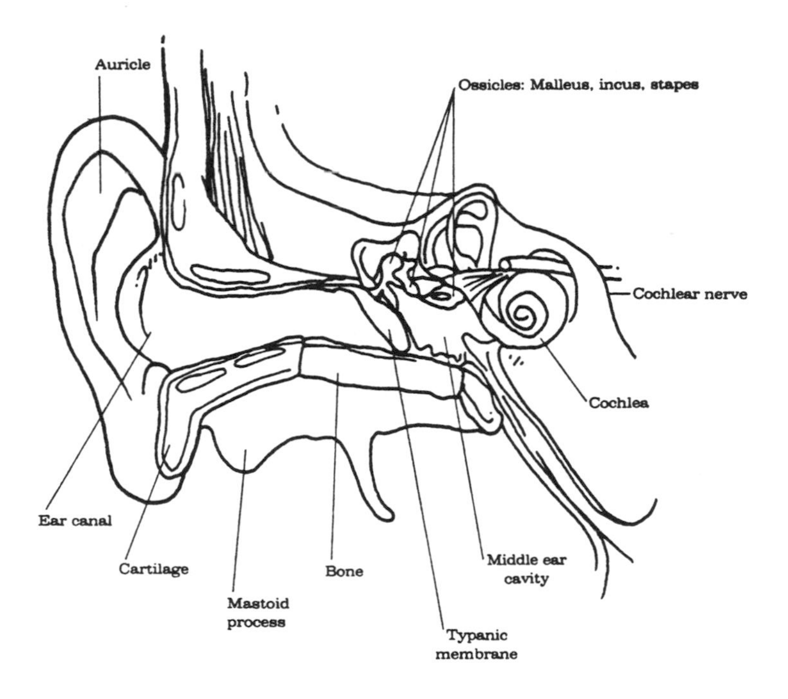

Figure 3C

The middle ear contains the ear drum and the three delicate ear bones: the malleus (hammer), incus (anvil) and stapes (stirrup). The middle ear functions include:

Transmission and amplification of sound waves
Protection of the inner ear from loud sounds
Maintenance of equal pressure within the inner ear

The inner ear has two major parts, the cochlea and the auditory nerve. Inner ear primary function includes:

The translation of vibration from sound waves into nerve impulses with subsequent transmission to the brain.

Eye

The eye is the vehicle for receiving visual stimulation. See Figure D for a drawing of the eye.

Parts of the Eye

Parts of the eye	Comment/Function the eye
Sclera	White outer covering of the eye Gives the eye its shape Protects the inner eye contents
Cornea	Covers the iris (cornea is transparent) Is bent to receive light rays Is visible to the naked eye from the side
Choroid	Lines the sclera's inner surface Absorbs light rays
Iris	Located between the cornea and the pupil Colored portion of eye Brown or dark iris have more pigment Blue iris' have no pigment Regulates the amount of light entering the eye Contracts with dim light allowing the pupil to receive more light
Pupil	Hole in the center of the iris; size can be regulated Light is transmitted via the pupil to the retina
Lens	Transparent disk behind the pupil Has the ability to change shape
Retina	Covers the choroid Primary function is image formation

The Eye

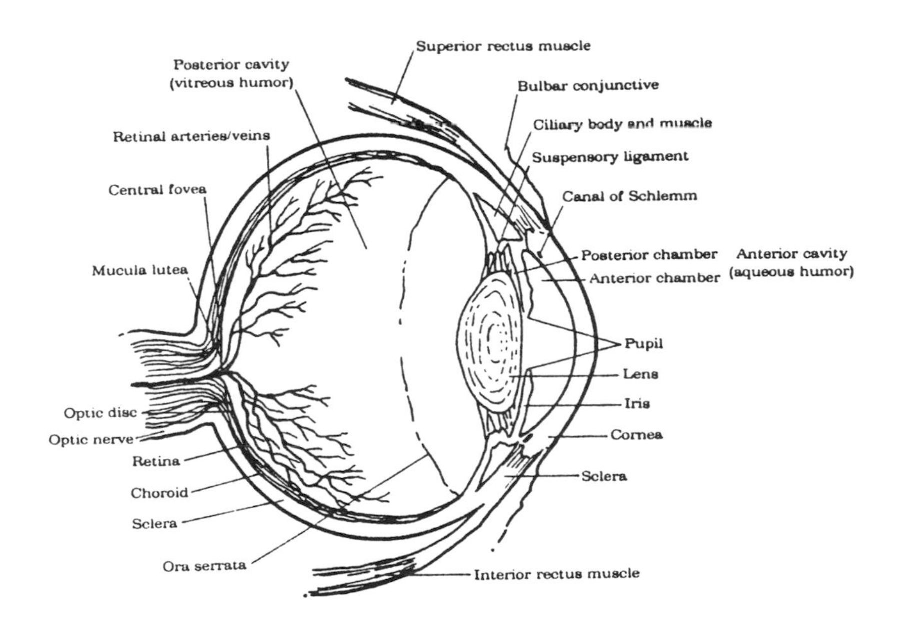

Figure 3D

Section II -- ASSESSMENT

Health history

Chief complaint

Examples of chief complaints:

Mental confusion, impaired memory or loss of mental abilities
Dizziness
Tinnitus (ringing sensation in ears)
Numbness or tingling in extremities
Unsteady gait
Tremors
Weakness
Hypotonia (loss of muscle tone)
Hypertonia (spasticity of muscles)
Double vision, impaired visual acuity
Pupil constriction or dilation
Difficulty in swallowing
Impaired speech
Seizures

Personal and family history

The family or personal history may supply clues that will aid in the diagnosis.

Psychiatric illness/migraines/multiple sclerosis

Some psychiatric illnesses, migraines, and multiple sclerosis have an increased family incidence suggesting that genetic or environmental factors may influence susceptibility.

Seizures

Determine onset date, type of seizure activity, if known (generalized, absence, tonic-clonic) and the frequency that seizures occur.

Head injury/spinal cord injury

For injuries determine the date, cause of injury, and any subsequent disabilities.

Headaches

Headaches may be caused by organic disease such as brain tumor, brain abscess, meningitis, hemorrhage, vascular changes, alcoholism, poisonings, or hypertension. Other causes include head injury or psychogenic causes such as anxiety, hysteria, or muscle tension. The location, severity, frequency, and duration combined with other elements of the history will aid in determining the cause.

Neurological system testing

A variety of tests may have been completed or ordered to include electroencephalography (EEG), spinal tap, cerebral angiogram, electromyography, myelogram, and skull x-rays. Scans of the brain include computerized tomography (CT), magnetic resonance imaging (MRI), and positron emission tomography (PET). Hearing and vision testing may include audiometic testing, brain stem audio evoked response (BAER), visual evoked response (VER) or simple screening procedures.

Some laboratory tests that may be ordered include cerebral spinal fluid analysis and therapeutic drug levels.

Current information

List all medications being taken and physicians the patient currently sees; ask if patient has seen a neurologist.

NEUROLOGIC SYSTEM ASSESSMENT FORM

Chief Complaint

Patient's statement____________________Onset______Symptoms__________________

Frequency____________Duration____________Other areas affected________________

Have you had this before?________________Date of last episode_________________

What treatment was given?____

What do you think caused this to happen?_________________________________

What changes have you made in your life because of this problem?

____

Personal and family history

	Patient	Family member		Patient
Psychiatric Illness	_____	_____	Seizures	_____________ onset__________ Type______________________________ Frequency__________________________
Migraines	_____	_____		
Multiple Sclerosis	_____	_____	Head injury	___________date_____________ Disability__________________________
Alzheimers	_____	_____	Spinal cord injury	______________________ Date____________disability______________

Do you have numbness or tingling sensations (location)?__________________________

Difficulty walking, talking, swallowing or chewing?_____________________________

Muscle weakness/spasm____________Tremors__________Visual problems____________

Headache (location, severity, frequency, duration)______________________________

Dizziness_______________Memory or concentration loss ___________Insomnia________

Neurological system testing

EEG____________Scan of brain (types)____________________Spinal tap______________

Cerebral angiogram____________Electromyography___________Myelogram__________

Hearing/vision testing_____________Skull x-ray__________Blood tests______________

Current treatments/medications

Who is your physician?_________________________________Phone_________________

Medications:

Name____________________________Dose___________Frequency_______Route_______

Name____________________________Dose___________Frequency_______Route_______

Is there anything else you want me to know?_______________________________

PHYSICAL ASSESSMENT

Physical assessment of the neurological system is conducted in the following order:

1. Inspection
2. Palpation

The assessment of this system begins with the health history assessment. Observe the patient carefully for signs of neurological deficits. Assess the ability to understand questions and to reply appropriately. Is the speech clear? Is the patient able to recall his/her medical history? Observe the motor status. Did any tremors occur? Are movements smooth and coordinated?

INSPECTION

General appearance/ affect

This includes posture, behavior, coordination, and grooming as well as the ability to interact with the examiner. Does the patient appear confident, depressed, lethargic or energetic?

Affect refers to the emotional state. Is the patient elated, happy, quiet, thoughtful, depressed, angry, or hostile?

Level of consciousness

There are several levels of consciousness:

1. Fully conscious

   The patient is alert and awake. Responds quickly and appropriately to verbal questions. Oriented to time, place, person and situation.
2. Confused

   Provides an inappropriate response to questions, decreased memory and attention span. May follow simple commands.
3. Lethargic

   Very drowsy affect with longer sleep periods noted. Responds slowly and appropriately after a delay. Difficult to awake, may return to sleep immediately.
4. Delirious

   Anxious with marked confusion. Perceptions are distorted, attention span is decreased. Reactions to stimulus are inappropriate.
5. Stuporous

   Early coma. Unconscious, may be aroused for short periods of time with intense stimulation. Able to follow some simple commands, responses are slow. Pupillary reflexes are sluggish.
6. Light coma

Moans in response to painful stimuli. Flexion motor response or mass movement.

7. Deep coma

   Decerebrate posturing to painful stimuli.

8. Deeper comatose state

   Muscles are flaccid. No pupillary reaction to light. No spontaneous respirations. Some deep tendon reflexes may be present. No response to stimuli.

9. Brain death

   Check agency guidelines, in general must meet these:

   Absence of brain waves on two EEG's (24 hours)
   No cerebral function detected
   Failure of cerebral perfusion
   Physician rules out hypothermia or drug toxicity

Speech quality

Includes receptive and expressive speech. Receptive speech is the ability to comprehend what is said (follow commands); expressive speech is the ability to communicate (speak, sign, gesture).

Orientation/memory/judgment

Orientation is the awareness of what is going on. Does the patient know name, date, where they are and why? If so, they are oriented X4. Memory questions can include immediate memory (repeating a sequence of numbers), recent memory (what was eaten for breakfast) or remote memory (mother's maiden name). Judgment questions are indicators of higher brain function; a sample question might be: "What do you if you are tired?"

Motor function

Is the posture erect or bent? Is the gait normal, halting, ataxic, or shuffling? To perform a Romberg test, have patient stand with feet together with eyes open and then closed. Stay near to protect patient if fall occurs. Only minimal body swaying should be present. Record, in seconds, how long the patient can stand on one foot.

Cranial nerve testing

The cranial nerves can be tested by using materials readily available; for example when testing CN I, can the patient with eyes closed identify coffee, alcohol, or soap? The six cardinal fields of vision can be tested by having the patient track a pen that makes an imaginary H in the air. Is the patient able to track with both eyes in all the fields? Note any nystagmus or paralysis. PERL stands for pupils equal and reactive to light. Check CN VII by having the patient smile, frown, and puff out his/her cheeks. CN VIII has been tested during the interview; can the patient hear you speak? Try a whisper while covering one ear (patient's) at a time, and record response. Test CN IX by having patient taste and identify sugar and salt. To test CN IX and X, observe uvula while patient says "ah"; is it symmetrical? Is a swallow and gag reflex present? To test CN XII, observe tongue in mouth; are fasciculations (fine tremors) present? Does the tongue deviate to one side or does it rest in midline?

PALPATION

Cranial nerve testing

To test CN V, have patient firmly clench teeth while you place your fingertips on his/her temples and then at the base of the jaw on the masseter muscle; feel for an asymmetry. To test CN XI, have patient raise shoulders while you try to push them down; is movement and strength equal? Test head movement by holding your hand to one side of the cheek and instruct patient to push with his/her head against it; observe movement and strength of sternomastoid on opposite side.

Deep tendon reflexes

Record response on figure provided using numeric guide.

4+	Hyperactive very brisk,
3+	Above average response but not abnormal
2+	Average
1+	Diminished to low normal
0	No response

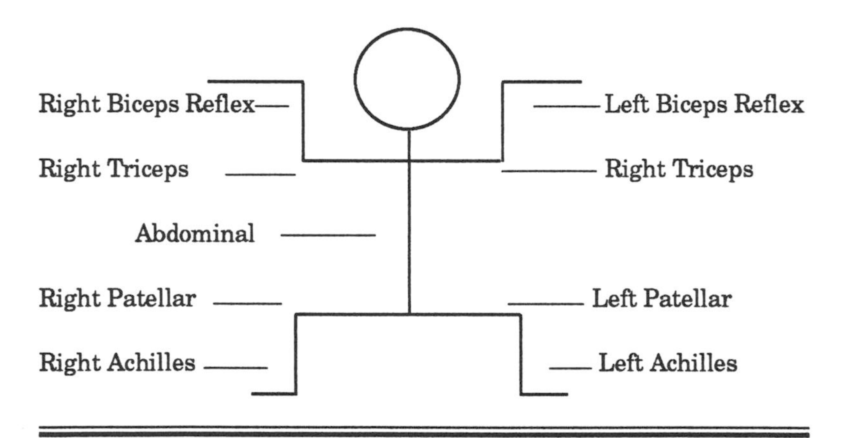

NEUROLOGIC SYSTEM PHYSICAL ASSESSMENT FORM

INSPECTION:

General appearance__________________________________Affect__________________

Level of Consciousness_________________________Speech quality__________________

Orientation (name/date/place/situation)____

Memory (address/president/state)___________________________Judgment____________

Motor function: (posture/gait/Romberg/one foot standing)___________________________

CN I: (Olfactory) Identify coffee_________________alcohol______________soap__________

CN II: (optic) read print from 12 inches (rt./lt.eye)__________________/______________

CN III (oculomotor) IV (trochlear) VI (abducens): fields of gaze

Rt.	Right eye	Lt.	Rt.	Left eye	Lt.
Up					Up
(CNIII)		(CN III)	(CN III)		(CN III)
CN VI)	(CN III)	(CN III)	(CN III)	(CN III)	(CN VI)
(CN III)		(CN IV)	(CN IV)		(CN III)
Down					Down

CN III: PERL_____________CN VII: (facial)movement_____________________________

CN VIII: (acoustic) hearing______________CN IX: (glossophyngeal) taste_______________

CN IX, X: (vagus) swallow____________symmetry palate/uvula__________Gag__________

CN XII: (hypoglossal) tongue position___________________fasiculations_______________

PALPATION

CN V: (trigeminal) temporal___________masseter____________equal sensation__________

CN XI: (spinal accessory) raise shoulders_________________head movement____________

Assessment Notes :

____

____

____

Deep Tendon Reflexes

Rt_______ ______ Lt.

_______ _____ ______

_______ ______

Section III – LABORATORY & DIAGNOSTIC TESTS

The following table contains some of the more common laboratory tests used to assess this system.

Cerebrospinal fluid analysis

Indications: Assess for any abnormality in the cerebrospinal fluid (CSF).

Regarding collection: CSF is withdrawn during a lumbar puncture performed by a physician. Fluid is collected in three test tubes with each containing about 3 ml of CSF. The tubes are immediately transported to the laboratory for analysis. A signed consent form is required. During the procedure the patient must lie very still in a fetal position with the back bowed.

Results of CSF analysis:

Expected	Other findings	Comments
Crystal clear colorless CSF	Red or pink CSF Xanthochromia CSF (yellow colored) Cloudy CSF	Traumatic lumbar puncture Subarachnoid hemorrhage Cerebral hemorrhage Old blood in CSF (hours to days old) previous subarachnoid hemorrhage previous cerebral hemorrhage Discoloration may remain for 3 weeks Infection meningitis neurosyphillis
CSF pressure (mmH_2O) Adult 75-175 Child 50-100	Pressure decrease	Dehyration Hypovolemia
	Pressure increase	Increased intracranial pressure meningitis or encephalitis subarachnoid hemorrhage brain tumor or abcess
Cell count (WBC's) 3 (cu mm mm) Adult 0-8 Child 0-8 Newborn 0-15 Premature infant 0-20	Increased count	Viral infection poliomyelitis Aseptic meningitis Syphilis of the central nervous system Multiple Sclerosis Brain tumor Brain abscess Subarahnoid hemorrhage

Protein (mg/dl) Adult 15-45 Child 14-45 Infants Premature <400 Newborn 30-200 1-6 mo. 30-100	Increased protein	Meningitis Guillain-Barre syndrome Subarachnoid hemorrhage Brain tumor or abscess Syphilis of the central nervous system Trauma or traumatic lumbar puncture with blood in CSF
Chloride (MEq/L) Adult 118-132 Child 120-128	Decreased level	Tuberculosis meningitis Bacterial meningitis IV saline or electrolyte infusions may cause an inaccurate result
Glucose (mg/dl) Adult 40-80 Child 35-75 Newborn 20-40	Decreased glucose	Purulent meningitis Fungi, protozoa, pyogenic bacteria Subarachnoid hemorrhage Lymphoma or leukemia
	Increased glucose	Trauma Diabetes mellitus
Lactic acid content 14mg/dl	Increased content	Hypoxic or ischemic cerebral injury Bacterial or fungal meningitis Brain tumor
CSF culture No organism found in CSF	Positive culture	Causative organism is identified Meningitis-bacterial pneumococcal meningococcal RH. influenzae streptococcal

Table of Common Diagnostic Tests

Test Name	Indications	Comments
Audiometric testing Hearing adequate for speech	Possible hearing loss	**Pre-procedure:** Assess ability of patient to understand and follow directions Earphones will deliver a series of tones at varying intensities Results are plotted on audiogram Hearing loss will be recorded in decibels (dB) **Post-procedure:** No restrictions
Brain stem auditory evoked response (BAER or ABR) Normal activity of CN VIII	Infants and others who are unable to respond to audio-tone testing Profound hearing loss Sensorineural hearing losses Suspected feigned hearing loss Coma-determine brain stem function	**Pre-procedure:** Assess ability to lie still Pre-medication may be ordered Earphones will be worn along with EEG leads Electrical activity in the brain will be recorded in response to sound Procedure takes about 45 minutes **Post-procedure:** No activity restrictions Assess for level of anesthesia as appropriate
Cerebral angiogram Normal cerebral blood flow	To detect brain disease: Cerebral aneurysm, cerebral tumor or thrombosis, hematomas, fistula plaques, or spasm To determine blood flow	**Pre-procedure:** Signed consent form is required Explain procedure to patient NPO for 8-12 hours before testing Record baseline vital signs Remove dentures, metallic jewerly Premedication will be ordered IV should be started before test Patient will lie supine Procedure may take 1-2 hours Assess for allergy to iodine Post-procedure: Assess site for bleeding, edema Monitor vital signs frequently Bedrest for 12-24 hours Assess pulses in extremities Assess for dysphagia, confusion, slurred speech or weakness

Computed tomographic x-ray (CT scan)	Anatomic deviation Hydrocephaly Cortical atrophy Brain tumor Brain abscess Trauma	**Pre-procedure:** Signed consent form is required Assess orders for with or without contrast dye Assess for allergy to iodine or seafood Weight will be needed for the calculation of dye dosage **Post-procedure:** Assess for allergic reaction to dye if contrast CT scan was done No activity restrictions noted
Electroencephalography (EEG) Normal tracing regular short wave	Seizure disorders Brain tumors Cerebralvasculer accidents (stroke) Head traum Mengingitis Determine cerebral death (flat EEG)	Pre-procedure: Assess orders for type of EEG: awake, drowsy, asleep, with stimuli, or combination Hair should be clean and free of oils or hairspray Coffee, tea, cola are restricted Follow orders regarding sleep May take from 1-2 hours Sedative may be ordered Post-procedure: Remove collodion from hair No activity restrictions
Magnetic resonance imaging (MRI) Normal brain structur	Brain tumors Brain anomalise Brain edema Cerebral contusion Hydrocephyaly Suspected CVA	**Pre-procedure:** Explain procedure to patient The head will be placed in a magnetic field and a radiopulse will be used to image brain Procedure carries no known risk to the patient Assess patient for metallic items (prosthetics, plates, or pins) If present MRI is contraindicated Have patient remove all jewerly, clothing with metal and credit cards prior to imaging Explain test takes 1-2 hours and that a stedy rhythmic pounding will be heard **Post-procedure:** No activity restrictions

Myelogram (x-ray of spinal cord with contrast dye or air) Normal spinal subarachnoid space	Herniated disks Spinal tumor Spinal injury	**Pre-procedure:** Signed consent form is required Lumbar puncture will be performed so contrast dye can be injected into subarachnoid space Assess for allergy to iodine and seafood Contraindicated in increased ICP NPO for 4-8 hours before testing Enema may be ordered before test Premedication will be ordered **Post-procedure:** Bedrest with body flat 6-8 hours Monitor vital signs frequently Increase fluid intake Assess for complications such as meningitis
Positron emission tomography (PET)	Research scanning device used to map brain function Presently not used in diagnosis	**Pre-procedure:** Radioactive glucose is given IV during procedure and intensity of intake in brain is recorded on a computer
Superconducting quantum interference device (SQUID)	Research scanning for brain mapping	Device is new research tool using magnetic fields to monitor neural activity Not yet in clinical use
Visual evoked responses (VER) Normal optic nerve activity	Assess visual acticity in small child or infants Assess optic nerve involvement in multiple sclerosis	**Pre-procedure:** Assess ability to lie still Premedication may be ordered EEG leads will be placed to assess electrical activity of brain in response to stimuli Procedure takes about 45 min. **Post-procedure:** No activity restrictions Assess for level of anesthesia as appropriate
X-ray of skull or spine Normal cranial structure	Head trauma Trauma to neck Brain tumor Congenital bone anomalies	**Pre-procedure:** Remove any hairpins, glasses, or dentures before testing Procedure may take 10-15 minutes followed by a 10-15 minute wait to be sure films are readable Assess for pregnancy in females **Post-procedure:** No activity restrictions

Section IV -- PROCEDURES AND CONDITIONS

Seizure Disorders

Seizure disorder, convulsive disorder, and epilepsy are all terms used to describe recurrent seizure activity . A seizure is a symptom, not a disease. Seizures, like fever or any other symptoms may result from a variety of health problems. Possible causes of seizure activity are listed. However, more than 50% of seizures have no identifiable cause (idopathic).

Causes of seizure activity

- Structural brain damage

  Structural brain damage may occur at any time of life. Some possible causes include:

  Congenital abnormalities of the brain.

  Brain tissue death -- Anoxia, compression of brain tissue with increased intracranial pressure, intraventricular bleeding, or cerebrovascular accident.

  Head injury -- Automobile accidents, falls, gunshot wounds; blows to the head or other cerebral trauma.

  Brain tumors -- Benign or malignant growths in brain tissue or metastatic tumors from lung, breast or other tissue which have spread to brain tissue, changing the structure.
- Degenerative brain disorders

  Generally associated with aging, however, degenerative brain disease and dementia may occur in younger patients.

  Dementia -- Progressive decline in intellectual function may be caused by Alzheimer's disease, Huntington's chorea, multiple sclerosis, or acquired immunodeficiency syndrome.
- Central nervous system infections

CNS infections may be caused by a variety of agents including:

Bacterial infections -- Acute bacterial meningitis may be caused by *Neisseria meningitidis, Hemophilus influenza,* or *Streptococcus pneumoniae.* Subacute meningitis may be caused by tuberculous meningitis.

Viral infections -- Acute viral encephalitis or aseptic meningitis may be caused by a variety of viruses to include *poliovirus, coxsackie virus,* or *herpes zoster.*

Fungal infections -- Subacute meningitis may be caused by *Cryptococcus, Candida,* or *Aspergillus.*

Acquired immunodeficiency syndrome (AIDS) -- Neurologic symptoms may be the first sign of AIDS. Degenerative brain disease may result from Toxoplasma encephalitis, cryptococcal and tuberculosis meningitides, multifocal leukoencephalopathy or other infections.

- Metabolic imbalances

  May occur at any time during life and cause a single seizure episode which usually disappears when the underlying problem is corrected. Metabolic imbalances may lead to onset of seizure disorder, if damage to brain tissues occurs.

  Oxygen deprivation -- Cerebral hypoxia due to respiratory or cardiac arrest, shock, anesthesia, carbon monoxide poisoning, or trauma.

  Hypovolemia -- Dehydration

  Increased temperature -- Febrile seizure in a child under 6 years

  Electrolyte, mineral, or acid-base imbalances

  Calcium imbalance, acidosis, hypoglycemia.

What is a Seizure?

A seizure is an unpredictable, abnormal surge of electrical activity within the brain that is transient. Unlike fever which can be defined, seizures do not have a single constant factor that can be always be recognized. The abnormal electrical activity may affect the entire body with loss of consciousness and tonic clonic movements, or it may only appear as lip smacking, ringing in the ears, or the purposeless movement of an extremity. A classification system was devised to group seizures into categories because of the variety of seizure manifestations. This system is not perfect and not all seizure activity fits into one of the categories. The type of seizure may change from one episode to another, or begin with one type and develop into another.

Classification of seizures

Seizures can be classified several different ways. One classification system divides seizures types into four groups:

I. Partial seizures -- seizures begin locally

Symptoms are dependent upon the specific lesion site.

A. Partial seizures with elementary symptomatology. Consciousness is usually maintained.
   1. With motor symptoms
      Localized twitching of muscles (Jacksonian)
      Motor symptoms may spread up extremity
      Jerking of one arm or leg
      Transient paralysis (minutes to hours) may follow

Lip smacking
2. With sensory or somatosensory symptoms
Distorted vision
Flashing lights
Ringing in the ears
Unusual sensations in one or more body parts
3. With autonomic symptoms
Sweating
Pupillary dilation
4. Compound forms

B. Partial seizures with complex symptomatology. Consciousness is usually impaired.
1. With impaired consciousness only
1-2 minute loss of contact with surroundings
2. With cognitive symptomatology
Mental confusion
Inability to understand verbalizations
May resist help
3. With affective symptomatology
Auditory or visual hallucinations
Illusions
4. Psychosensory symptomtology
Intense emotions may occur (anger, fear)
Forced thinking may occur
5. Psychomotor symptomatology
Continuation of an activity begun prior to seizure
Verbalizations are unintelligible

C. Partial seizures leading to generalized seizures

II. Generalized Seizures

EEG taken during a seizure will show generalized activity that is synchronized bilaterally.

1. Absences (petit mal)
Momentary loss of consciousness
Brief loss of postural tone
Lasts less than 10 seconds
No aura is present
Sudden return to full consciousness
2. Bilateral massive epileptic myoclonus
Consciousness may be altered or maintained
Rapid muscle contractions occur
Single or repetitive brief myoclonic jerks
Balance may be lost
3. Infantile spasms
Usually onset occurs at 6-12 months of age
Sudden flexion of head and trunk
May have from few to hundreds of episodes daily

4. Clonic seizures
   Clonic jerking only occurs
   Rhythmic jerking of extremities
5. Tonic seizures
   Tonic contraction only occurs
   Consists of stiffening of entire body
   Cyanosis may occur
6. Tonic-clonic seizures (grand mal)
   Classic epileptic attack
   Epileptic cry may be present
   Pupils may dilate and deviate upward
   Tonic and clonic motions occur
   Tongue or lips may be bitten
   Urinary or fecal incontinence may occur
   Usually lasts 2-4 minutes
   Postictally a deep sleep may follow
7. Atonic seizures
   Sudden and usually complete loss of muscle tone in limb, neck, and trunk muscles
   No warning before, therefore, injury may occur
8. Akinetic seizures
   Arrest of movement without a significant loss of muscle tone
   Consciousness is lost
   Full awareness returns quickly

III. Unilateral seizures

Involving only one side or predominantly one side of the body

IV. Unclassified epileptic seizures

Not enough data available to classify

Interventions for patients with seizure activity

Obtain history:

Obtain a complete seizure history from patient or significant other (patient may have no memory of activity.)

Family history -- Assess for other family members with seizure disorder or with neurological diseases.

Birth history -- Assess for any congenital birth defects, anoxia during labor or delivery, or neonatal seizure onset. If seizures began in the neonatal period assess developmental history for delays.

Onset of seizures -- Age at onset, any illness or trauma that preceded seizure onset such as automobile accident, near drowning, or cerebral infections.

Frequency of seizures -- Frequency at onset, current frequency of seizures. Date of last seizure activity.

Type(s) of seizure(s) if known -- Assess knowledge of seizure type. Have significant other describe last seizure including how long the seizure lasted, if consciousness was lost and for how long, any motor movements (stiffening, jerking, loss of movement), pupillary reaction, urinary or fecal incontinence, or behavior changes. Ask if last seizure was typical of most of the seizures patient has.

Seizure auras or triggers -- Assess if patient can predict the seizure by recognizing a sensation (aura) and assess how much warning patient has of impending activity. Is there anything that the patient or family knows triggers that seizure activity? Some seizures are triggered that by the following stimuli.

Loss of sleep	Menstrual period	Missed medication
Illness	High fever	Loud noise
Emotional stress	Flashing lights	Missed meal

Medication history -- What medications have been taken in the past, what is the current medication(s); include name, route, dosage, and frequency taken of each medication. Are the seizures controlled with medication?

Postictally (following the seizure) -- Assess the length of time before full consciousness is returned. Assess for muscular aches, weakness, paralysis or headaches following seizures.

Institute seizure precautions:

Padded side rails which are kept up at all times to prevent accidental fall and injury, if seizure occurs.

Oral airway kept at bedside

Frequent checks on patient's condition

Administer anticonvulsants and monitor laboratory work

- Assess orders for anticonvulsant therapy
- Monitor seizure medications/serum drug levels and report
- Monitor related laboratory studies and report abnormal values
- Monitor for and report side effects from seizure medications

Perform patient teaching as needed

- Provide teaching regarding basic knowledge of seizures
  Definition of seizures and or seizure disorder
  Identification of stimuli that may trigger a seizure
- Provide teaching on medication administration
  Instruct patient in importance of compliance with medication regime. Noncompliance is the primary cause of status epilepticus, a life threatening condition where seizure activity continues for greater than 30 minutes.
- Provide family with instruction on what to do for seizures

Interventions to be used during seizure activity

- Remain calm
- Move any objects away that may cause injury
- Do not attempt to restrain the patient as this may cause broken bones
- Do not attempt to insert an airway, if teeth are clenched. If teeth are not clenched, place oral airway or soft washcloth between teeth
- Turn patient's head to side to prevent aspiration of excess saliva
- Place soft pad under patient's head
- Loosen tight clothing
- Remain with the patient until seizure activity stops and consciousness is regained
- Observe characteristics of seizure to include motor activity, pupillary changes, urinary or fecal incontinence, behavior changes, respiratory changes, pallor. Note time of onset as well as time seizure terminates
- Call for help, if seizure lasts more than 3-5 minutes
- Assess vital signs and perform neurological assessment
- Reassure and offer emotional support to patient and family
- Call physician and reportseizure activity
- Chart all observations from before, during, and after seizure activity

Coma

Definition

A prolonged impairment of consciousness where the patient is unarousable and unresponsive in an abnormally deep sleep. Coma may be due to illness or injury.

Causes of Coma

Like a seizure, a coma is a symptom, not a disease.

Coma can be due to a variety of causes including:

- Traumatic head injury
  - Car accidents
  - Falls
  - Gunshot wounds
- Circulatory disorders within the brain
  - Cerebral hypertension
  - Cerebrovascular accidents (CVA)
  - Increased intracranial pressure
  - Insufficient blood flow to brain tissues
  - Intracerebral hemorrhage
  - Subarachnoid hemorrhage
  - Thrombus formation
  - Uncontrolled seizures
- Infections
  - Acute botulism
  - Brain abscesses
  - Encephalitis
  - Meningitis
- Drug overdoses
  - Alcohol
  - Barbiturate
  - Insulin
- Metabolic Causes
  - Acid-based imbalances
  - Diabetic coma (lack of insulin)
  - Hepatic disease
  - Hypothermia
  - Uremic coma

Assessment of coma:

Glasgow coma scale

The Glasgow coma scale is one of several tools used to assess level of consciousness. The higher the score, the more responsive the patient is. A low score, for example, 7, indicates the patient is not responsive.

GLASCOW COMA SCALE

EYES OPEN	
Spontaneously	4
To speech	3
To pain	2
No response	1
VERBAL RESPONSE	
Oriented/talking	5
Disoriented/talking	4
Words inappropriate	3
Incomprehensible	2
No response	1
MOTOR RESPONSE	
Obeys command	6
Localizes pain	5
Flexion: Withdrawal	4
Decorticate: Flexion	3
Decerebrate: Rigidity	2
No response	1

In addition to the Glasgow coma scale, the neurological assessment also includes an evaluation of pupil response and muscular strength.

Pupillary response

- Pupils normally contract in strong light (such as a penlight) and dilate in dim light. Pupil size may range from 2-6mm.

- Darken room prior to the examination of the pupils
- Pupil reaction time is normally brisk, examples of reaction time choices are: brisk, slow, or fixed (unresponsive)
- Pupils should react equally to light

- Direct pupil reaction is the constriction of the pupil in the same eye that the light beam strikes
- Consensual pupil reaction is the constriction of the pupil in the opposite eye that the light beam strikes
- Record size and reaction time of each pupil

Muscle strength

Degree of muscle strength is also an indicator of neurological status.

- Muscle tone and muscle strength should be equal bilaterally in both the upper and lower extremities
- Muscle strength in upper extremities can be tested by having the patient grasp and squeeze one of your fingers. Record grasp as absent, weak, moderate, or strong. Test both upper extremities
- Muscle strength in the lower extremities can be tested by having the patient push with his/her leg while the foot rests against your hand. Record response as absent, weak, moderate, or strong. Test both lower extremities

Neurological assessments are usually repeated every 1-4 hrs. to assess for changes.

Decorticate and decerebrate posturing describe postures which may be seen in comatose patients.

Decorticate - Adduction of arm
Flexion of elbow
Flexion of wrist

Decerebrate - Adduction of arm
Internal rotation and extension of arm

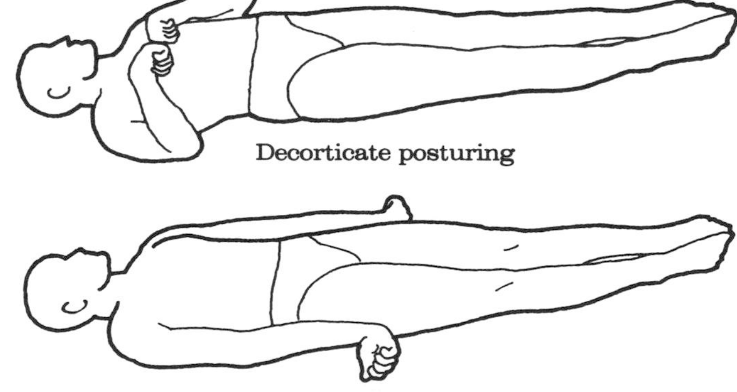

Decorticate posturing

Decrebrate posturing

Care for the comatose patient

The patient who is in a coma is totally dependent upon others for care and is unable to meet their most basic needs. Needs should be prioritized.

Need	Interventions
Airway	Place and maintain oropharyngeal airway if not already in place
	Position patient on side to prevent aspiration of secretions
	Suction frequently to remove secretions which may block the airway
	Assist as needed in placement of endotracheal (ET) tube
	Restrain arms if needed to prevent accidental dislodgement of oropharyngeal or ET airway device
Breathing	Assess baseline respiratory rate and rhythm
	If spontaneous breathing is not present, provide oxygenation via rescue breathing or ambu bag to maintain oxygenation until a ventilator can be obtained
	Provide oxygen as needed and ordered
	If on ventilator, assess orders for ventilator settings and compare with actual settings. Assess level of ET tube and check that tubing is secured
	Check all alarms for function, be sure alarms are on
	Suction as needed to maintain patency of airway
	Monitor respiratory rate and rhythm on regular basis
	Monitor pulse oxyimeter and arterial blood gas values to assess oxygenation status
	Assess pattern of respirations and record
	Cheyne-Stokes
	Rapid deep breathing alternating with slow shallow breathing
	Hyperventilation
	Rapid deep breathing

Circulation	Monitor pulse, blood pressure and temperature on regular basis even patient is on monitor. Monitor reading may be inaccurate
	Connect patient to cardiac monitor and assess for heart arrhythmias
	Monitor laboratory values and report any abnormal values
	Administer medications as ordered to promote cardiac output/ circulation
	Maintain intravenous catheter to provide route for fluids and medications as ordered
	Establish and maintain urinary catheter to assess renal blood flow and monitor output
	Monitor all intake and output for any imbalances
Nourishment	Maintain nasogastric, gastrostomy tube, or total parental nutrition as ordered
	Assess weight at least every other day and record
	Consult with physician and dietitian if weight loss is uncontrolled
	Assess frequency and character of stool
	Activities of daily living
	Reposition patient every two hours to prevent skin breakdown
	Provide range of motion exercises regularly to increase blood circulation and prevent contractures of joints
	Provide meticulous skin care to prevent skin deterioration
	Provide meticulous oral care to prevent murosal deterioration and reduce halitosis
Sensory stimulation	Talk to patient frequently, even if totally unresponsive, to provide auditory stimulation
	Touch the patient often during care to provide tactile stimulation
	Play music or taped recording of family and friends to provide auditory stimulation
	Place brightly colored posters in room, hang mobiles from ceiling to provide visual stimulation

Increased Intracranial Pressure (ICP)

ICP is the pressure the cerebrospinal fluid exerts within the ventricles of the brain. Normal ICP is between 0-15 mm Hg. When the ICP rises above 15 mm Hg, it is considered elevated and becomes a life-threatening condition. Normally tissue composes about 80% of the brain contents; the cerebrospinal fluid makes up about 10%, and the remaining 10% is blood. If any component within the brain increases, the pressure increases because the adult cranium cannot expand. An example of this is cerebral edema. Initially, in cerebral edema, the blood flow will decrease to compensate so that the presssure will remain about the same. If the fluid continues to accumulate and blood flow continues to decrease, the brain tissue becomes hypoxic. When cerebral hypoxia occurs, vasodilation results which increases blood flow, and therefore, volume. The increased blood volume and swollen tisssue now cause the intracranial pressure to increase. If the pressure is not relieved, a herniation of the brain will result and death will occur. The type of herniation depends on the basic problem and its location.

Causes of ICP include:

I. Increased fluid accumulation or tissue mass in brain

A. Brain tumor -- Growth of an intracranial neoplasm causes destruction and compression of normal brain tissue leading to cerebral edema and ICP.

B. Brain trauma -- When the brain is injured, it swells (cerebral edema) taking up more room within the skull. Trauma may be the result of automobile accident, fall, or blow to the head.

C. Cerebrovascular accident (CVA) -- A CVA may lead to brain tissue hypoxia and cerebral edema secondary to insufficient blood flow to the brain tissue such as occurs with embolism or thrombosis of the cerebral arteries.

II. Increase in cerebrospinal fluid

A. Hydrocephaly -- Obstruction of CSF or failure of the body to reabsorb CSF, allowing for a collection within the ventricles. The increased fluid increases the pressure.

III. Increase in cerebral blood volume

A. Brain trauma -- Trauma to head may rupture or injure cerebral vessels leading to an accumulation of blood within the skull which increases intracranial pressure.

B. Cerebrovascular accident (CVA) -- CVA may result from cerebral hemorrhage, cerebral arteriovenous malformation or aneurysm.

Signs and symptoms of ICP

Early assessment of ICP is critical to prevent brain damage. This assessment guide indicates the most common signs and symptoms of ICP. The severity of symptoms will depend on how high the ICP is and which area of the brain is being compressed. The best indicator of ICP is direct measurement via intraventricular catheter, subdural bolt or epidural sensor.

1. Assess for a decrease in level of consciousness (LOC)

   One of the earliest signs of neurologic deterioration.
   Assess for patient level of arousal (alert, lethargic, stuporous, comatose, ect.). Early in ICP, restlessness, or lethargy may be present. In later increased ICP; coma will be present.
   Assess for orientation to time, place, person and situation. Disorientation often occurs first to time, then to place. Disorientation to person occurs late in ICP.
   Changes in the LOC occur due to compression of the brain tissue.

2. Assess for pupillary function

   PERL (pupils equally reactive to light)
   Check both pupils during each assessment to see if they constrict in response to light. Check for consensual response.
   In ICP, the third cranial nerve may be compressed if the brain begins to herniate. If compression has occurred, the pupil will remain dilated in response to light or the pupillary reaction will be very slow. The pupil will also appear larger than the other pupil. As the ICP rises, you will see bilateral pupillary dilation and fixation.

3. Assess for visual disturbances

   Assess by having the patient track an object through the six cardinal fields of gaze as well as reading something. Be alert for complaints of visual problems.
   Diplopia (double vision), blurred vision, or loss of visual acuity occurs with increased ICP when pressure is placed on the cranial nerves that control eye movement.

4. Assess motor function

   Assess the ability to move all extremities on command. Assess strength of handgrip and legs.
   Muscle weakness develops when pressure is placed on the cortical tracts of the brain by compression. The weakness on one side of the body will reflect brain compression on the opposite side. In late stage increased ICP, decorticate or decerebrate posturing may occur.

5. Assess for headache

   Headache is not always present with ICP. At the onset headache may be vague; later as ICP rises, headache will be more severe especially with coughing, straining or bending over.

6. Vomiting

   Assess if projectile and the frequency of episodes. Assess if the patient was nauseated before the onset.
   Due to compression of the vomiting regulatory center, may be projectile and occur without nausea.

7. Assess vital signs

   Early in ICP, vital signs should be within normal limits.
   As the ICP exerts pressure on the brain stem, the vital signs will change. Bradycardia and a widening pulse pressure may occur (Cushing's response). Cushing's response is a late sign of ICP. The blood pressure will decrease sharply.
   Slowing of the respiratory rate occurs early. Later Cheyne-Stokes respirations may occur or ataxic breathing; these are related to brain stem compression.

8. Assess temperature

   Assess for increased rectal temperature.
   Temperature increases as the hypothalamus is compressed due to cerebral edema. This occurs in late ICP.

9. Assess for seizures

   Seizures may occur due to abnormal electrical activity within the brain with compression of tissues.

Section V– DRUGS

This table supplies only general information, a drug handbook should be consulted for details.

ANALGESICS

Action: To relieve pain

Indications: Pain/discomfort related to a wide variety of causes to include: headache, toothache, muscular, orthopedic, or abdominal pain, surgical procedures, burns, also used in trauma, childbirth labor, and terminal conditions.

General comments: Analgesics may be classified as narcotic or non-narcotics.

Narcotic Analgesics

- For moderate to severe pain
- Generally used for acute pain
- Post-operative, trauma, coronary disease
- Used for intractable pain experienced in terminal illness
- Respiratory depression may occur and respirations should be taken prior to use, hold if respirations less than 12
- Euphoria is a side effect and may contribute to psychological/physical dependence
- Sedation/decreased mental alertness may occur. These medications should be administered with patient instruction to avoid getting out of bed without assistance, or any hazardous activities (driving or operating equipment)
- Concurrent use of alcohol or other CNS depressants should be avoided due to potential for respiratory depression
- Generally IV dose is small and should be administered slowly to prevent respiratory depression
- Require a prescription to obtain medication

Non-narcotic Analgesics

- For mild to moderate pain
- May be used for acute or chronic pain; toothache, headache, malaise due to viruses, minor injuries, rheumatic joint complaints
- These are available in many combinations with aspirin, acetaminophen, caffeine common ingredients

- Over the counter preparations may be abused. Instruct to take dose as recommended by physician or per package directions.

Examples of drugs used in this classification:

Generic	Trade	Comments
Acetaminophen	Tylenol Tempra Datril	Non-narcotic analgesic Given PO or rectal Tablets may be crushed For mild to moderate pain Avoid use in alcoholics due to hepatotoxcity Also an antipyretic
Aspirin	Bayer Empirin Aspergum	Non-narcotic analgesics Given PO, rectal Administer with food, milk or water to decrease GI upset Tablets may be crushed Discard tablets if vinegar odor is present For mild to moderate pain Prophylacticuse to prevent heart attack or stroke Effective for rheumatic joint diseases, as an anti-inflammatory Tinnitis may indicate toxicity Not recommended for children, due to association with Reye's syndrome
Codeine	Methylmorphine Tylenol #3	Narcotic analgesic Given PO, SC, IM Administer with milk or food to prevent GI upset For moderate pain
Hydromorphone	Dilaudid	Narcotic analgesic Given PO, Rectal, SC, IM, IV slow For moderate to severe pain 8-10 times more powerful than morphine
Ibuprofen Ibuprin	Advil Motrin Nuprin Pamprin-IB	Non-narcotic analgesic Given PO for mild to moderate pain Nonsteroid antiinflammatory agent Tablet may be crushed
Meperidine	Demerol	Narcotic analgesic Given PO, SC, IM, IV slow For moderate to severe pain

Morphine sulfate		Narcotic analgesic Given PO, Rectal, SC, IM, IV slow Urinary retention may occur, encourage voiding
Pentazocine	Talwin	Narcotic analgesic Given PO, SC, IM, IV For moderate to severe pain
Propoxyphene	Darvon Napsylate	Given PO Narcotic analgesic Capsules may be opened and mixed with food For mild to moderate pain Smoking decreases effectiveness of this drug

ANTICONVULSANTS

Action: To prevent seizure activity.

Indications: Potential for seizures from any cause to include epilepsy, head injury, neural surgery.

General comments: Seizure medications are generally prescribed according to the type of seizure activity noted. Drowsiness is a possible side effect in all of the following medications and patients should be cautioned against activities that could be hazardous including driving until exact effect of medication is known. All seizure medications must be taken exactly as ordered, missed doses may precipitate seizure activity with some medications. Medication should never be discontinued without the approval of a physician, and then it may require several weeks or months to wean the patient off some of these medications. Regular laboratory monitoring is required. All persons taking anticonvulsants should wear a Medic-Alert bracelet at all times. No over-the-counter medications should be taken without consulting the primary health care provider.

Examples of drugs in this classification:

Generic	Trade	Comments:
Carbamazepine	Tegretol	Given PO Administer with meals or food Therapeutic level 4-12mcg/ml Avoid excessive exposure to sunlight due to photosensitivity Also used in treatment of trigeminal neuralgias, multiple sclerosis

Clonazepam	Klonopin	Given PO Long term use may require dosage adjustment to control seizure activity Monitor I&O, as drug may affect renal function
Ethosuximide	Zarontin	Given PO Therapeutic level is 40-100mcg/ml Store in light resistant container
Phenobarbital	Barbita	Given PO, rectal, SQ, IM, IV Tablets may be crushed if mixed with food or drink Administer IV cautiously, no more than 60 mg/minute Check IV catheter placement prior to administration, extravasation may require skin graft Physical dependence may occur Therapeutic level is 10-20mcg/ml
Phenytoin	Dilantin	Given PO, IV Check suspension strength carefully prior to administration for correct strength Shake suspension well Tablet may be crushed and mixed with food or fluid Administer with food or fluid Do not administer IV with any other drug Flush line with NS before and after IV administration Therapeutic level is 10-20 mcg/ml Assess for jaundice and report if found
Primidone	Mysoline	Given PO Tablet may be crushed and mixed with food or drink Therapeutic level is 5-10 mcg/ml
Valproic acid	Depakene	Given PO Do not crush tablets Oral solution should not be taken with soft drinks Therapeutic level is 50-100 mcg/ml

ANTIPSYCHOTICS (Neuroleptics)

Action: To decrease or eliminate psychotic behavior.

Indications: Used in the treatment of psychotic disorders to include Manic-depressive disorders, schizophrenia.

General comments: Long term therapy is usually required. Instruct the patient to continue medication until stopped by physician even if visible improvement is not seen or patient feels they no longer need the medication. Concurrent use of alcohol or other CNS medications should be avoided. Abrupt withdrawal from these medications is not advisable. Common side effects include:

- Discoloration of urine
- Sensitivity to sunlight, need for sunscreen
- Hypotension may occur after medication administration
- Extrapyramidal symptoms may occur to include pseudoparkinsonism, akathisia, tardive dyskinesia

Examples of drugs in this classification:

Generic	Trade	Comments:
Chlorpromazine	Thorazine	Given PO, rectal, IM, IV Tablet may be crushed Capsules should be swallowed whole IM injection should be given slowly and deep Inability to sweat may occur
Clozapine	Clozaril	Given PO May suppress bone marrow function May cause seizures to occur Patient must be monitored carefully
Haloperidol	Haldol Peridol	Given PO or IM Should be injected deep IM
Molindone	Moban	Given PO hydrochloride Xerostomia (dry mouth) may occur
Prochlorperazine	Compazine	Given PO, rectal, IM, IV Used also as an antiemetic
Promazine	Prozine Hydrochloride	Given PO (preferred), IM, IV Do not administer with antacids Liquid should be diluted with fluids IM injection should be deep
Thioridazine	Mellaril Thioril Millazine	Given PO Tablet may be crushed Take 1 hour before or after antacid Urine may turn pink-red to brown

Trifluoperazine	Stelazine Hydrochloride	Given PO, IM Do not give with antacids (within 2 days) Dilute oral concentrate with fluid Tablets may be crushed May have lower tolerance to cold

HYPNOTICS/SEDATIVES

Action: Hypnotics promote sleep. Sedatives promote relaxation and provide a calming effect

Indications: Hypnotics may be used when the promotion of sleep is the goal. Uses include insomnia, and sleep promotion prior to surgical procedures. Sedatives may be used to treat anxiety, acute reactions to stressful situations (grief reactions), agitation and "nerves".

General comments: The potential for dependence is high and the use of these medications should be monitored closely. Tolerance to drugs may develop requiring higher doses to achieve the same effect. Instruct patients not to share medication with others. Medication should not be administered to relieve pain. Alcohol and other medications that depress the CNS should be avoided.

Examples of medications in this classification:

Generic	Trade	Comments:
Amobarbital	Amytal	Given PO, IM, IV Sedative and hypnotic May be given in status epilepticus
Chloral hydrate	Noctec	Given PO, rectal Often used as a sedative/hypnotic in children and elderly Administer after meals to prevent GI upset
Flurazepam	Dalmane	Given PO Capsules may be opened and contents mixed with food or fluids

Paraldehyde	Paral	Given PO, rectal, IM, IV May be used in alcohol withdrawal for sedation Reported to have unpleasant taste Administer in juice or milk to improve taste and help prevent GI upset Discard unused portion of unit container after 24 hours Do not use if vinegar odor is noted Inject IM deep, in UOQ away from nerves as paralysis may result if nerve is injected
Phenobarbital		See anticonvulsants
Secobarbital	Seconal	Given PO, rectal, IM, IV May be given as a premedication prior to spinal or regional anesthesia
Temazepam	Restoril	Given PO Used primarily for early morning awakening, not for difficulty in falling asleep May aggravate confusion in elderly
Triazolam	Halcion	Given PO, Usually no daytime effects are present Used for short term insomnia Smoking decreases effectiveness Memory loss associated with drug

Section VI -- GLOSSARY

Acuity (vision) loss	Inability to see clearly; objects are blurred
Akathisia	Strong desire to move or an inability to sit. May be a side-effect of antipsychotic medications
Amblyopia "lazy eye"	Dim vision in one eye with no apparent anatomical cause
Anticonvulsant	Medication to prevent convulsions
Aphasia	Inability to understand spoken messages (receptive language) or communicate by speaking (expressive language); may be due to CVA or brain injury
Ataxia	Incoordination of muscle movements
Autonomic	One of the major nervous systems; it is divided into the sympathetic and parasympathetic systems
Aura	Subjective warning prior to a seizure
Brain abscess	Pus collected within the brain's tissue
Cerebrovascular (CVA) "Stroke"	Condition generally caused by a blockage. Accident in or rupture of a vessel in the brain
CNS	Central Nervous System
Conductive hearing loss	Sound waves are blocked from reaching the inner ear
Cortical vision loss	Brain is unable to interpret nerve impulses from the eye into a meaningful message
Diplopia	Double vision
EEG	Electroencephalography, recording brain electrical activity

Encephalitis	Inflammation of the brain
Encephalomyelitis	Inflammation of the brain and spinal cord
Epileptic cry	Sound which may occur with the onset of a tonic-clonic seizure, due to air being forced out of the vocal cords with the first contraction of the body's muscles
Epilepsy	Term for seizure disorder
Euphoria	A feeling of well being
Gingival hyperplasia	Swollen, enlarged, tender gums. May be side-effect of Dilantin
Hallucination	A visual, auditory or olfactory experience that has no basis in reality. A false sensation without external stimulation
Intraventricular hemorrhage	Bleeding within a ventricle in the brain
Meninges	The three membranes that cover the brain and spinal cord; the dura mater, arachnoid and pia mater
Meningitis	Inflammation of the meninges
Myopia	Near-sighted, able to see objects close up better than objects far away
Olfactory	Pertaining to the sense of smell
Pseudoparkinsonism	A combination of side effects that are similar to Parkinson's disease; these include a mask like face, arm tremors, and pill rolling movements
Status epilepticus	A life threatening condition where seizure activity continues for more than 30 minutes, or patient has one seizure after another
Tinnitus	Ringing in the ears

Hematolymphatic System

HEMATOLYMPHATIC SYSTEM

Table of contents

Section I -- OVERVIEW

Primary functions

1. The manufacture and development of blood cells
   a. Red Blood Cells (RBCs or erythrocytes) which:
   - Carry oxygen from the lungs to individual cells
   - Carry carbon dioxide from the individual cells to lungs

   b. White Blood Cells (WBCs or leukocytes) which:
   - Combat toxins and infection from bacteria and viruses
   - Induce inflammation to destroy foreign invaders

   c. Platelets (thrombocytes) which:
   - Prevent excessive blood loss via clotting

2. The filtering of blood to remove old blood cells, as well as bacteria, viruses, and toxins

3. Transport of nutrients, minerals, enzymes, hormones, vitamins and antibodies to the cells where needed

4. Transport of waste products from cells to the kidneys, sweat glands and lungs for removal from the body

Components and function

The hematolymphatic system is composed of bone marrow, RBCs, platelets, the liver, WBC's, plasma, lymph, lymph vessels, lymph nodes, the spleen, and the thymus

Bone marrow

Hematopoiesis (formation of blood cells) occurs in the red bone marrow. This includes the production of RBCs, WBCs, and platelets

Red Blood Cells (RBCs)

RBCs are transporter blood cells that contain millions of hemoglobin molecules. Each hemoglobin has four atoms of iron (heme) which can each hold one oxygen molecule

Heme picks up oxygen in the lungs and transports it to body tissues

The globin picks up carbon dioxide from the body's cells and transports it to the lungs for removal via exhalation

RBCs have antigens on the surface that determine blood type: A, B, AB, and O (absence of antigens)

O is the universal donor, AB is the universal recipient

Platelets

Platelets are fragments formed in the bone marrow and are important in hemostasis and blood clot formation

They are required to prevent severe hemorrhage.

Liver

Largest organ in the body and one of the most versatile: involved in many body functions

Important in the hematolymphatic system because it manufactures clotting factors

During fibrin clot formation as many as thirty different substances are needed to promote or prevent clot formation

Some of the important clotting factors formed in the liver are Factors V,VII, IX,X, and prothrombin

White Blood Cells (WBCs)

There are several types of WBCs and they have a nucleus (RBCs and platelets do not)

The primary function of WBCs is to fight infectious agents invading the body

Plasma

Liquid part of lymph fluid and blood (blood cells and platelets float in plasma). Plasma consists of electrolytes, glucose, proteins (albumin and globulin), nonprotein nitrogenous compounds, fats, bilirubin and gases

Provides a medium for the circulation of blood cells

Prevents leakage of fluids out of capillaries (albumin)

Carries nutrients to the cells and removes waste products

Components of plasma are important in maintaining balance in the blood: bicarbonate, carbon dioxide, chloride, phosphate and ammonia

Transports minerals, hormones, vitamins and antibodies

Lymph

Clear, transparent, or slightly yellow fluid formed in tissue spaces throughout the body and transported by lymph vessels

Composed of water, protein, fats and the end products of cell metabolism

Lymph derived from different parts of the body may have a slightly different composition

Lymph vessels

Carry lymph, in a one-way direction, into the venous blood system

Consists of small lymphatic capillaries and larger lymphatic vessels which have valves

The walls of lymphatic vessels are thinner the blood vessels

Lymph nodes

Found along the lymph vessels; these are encapsulated areas of lymphatic tissue where bacteria and foreign cells are filtered out of the lymph fluid

New lymphocytes are formed in the lymph nodes

Spleen

A dark red organ in the upper left abdomen just below the stomach

Forms all of the fetal blood cells

Forms lymphocytes from bone marrow precursors in the adult

Reservoir for blood storage

Filters blood; removal of bacteria and old blood cells

Thymus

Located between the sternum and the great vessels of the heart

Essential for the maturation of T-cells needed for the immune response

Erythropoiesis (production of RBCs)

The production of all blood is called hematopoiesis, the production of RBCs is erythropoiesis. Figure 4-1 shows the life cycle of a RBC from erythropoiesis to destruction. Red bone marrow is responsible for RBC production. Erythropoiesis is important to understand from the clinical viewpoint when assessing laboratory values. Reticulocyte counts may be ordered to assess the number of maturing RBCs. Normally only about 1% of all RBCs should be reticulocytes. If the count is very high it indicates an increase in RBC production which may occur when RBCs are being destroyed (hemolytic anemia), or lost (hemorrhage). The count also may be high when anemia has been corrected, and the body is working to replace RBCs. A very low reticulocyte count indicates bone marrow dysfunction.

LIFE CYCLE OF A RED BLOOD CELL

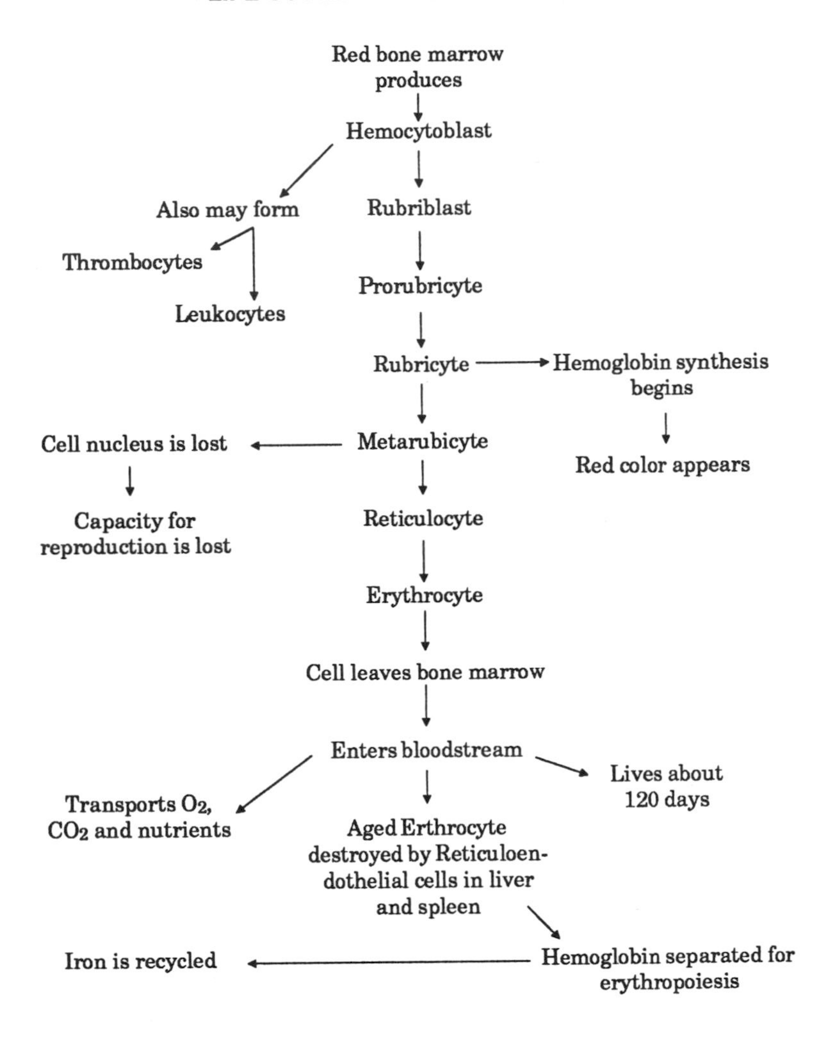

Figure 4-1

Section II -- ASSESSMENT

Assessment includes signs of immune system activation or failure as well as bleeding disorders or blood component abnormalities.

Health History

Chief complaint

Examples of common chief complaints include:

Excessive tiredness, lack of energy
Weakness, inability to do activities
Bleeding easily; nosebleeds, rectal bleeding, bleeding gums
Fainting or dizziness
Night sweats
Fever
Swollen glands
Frequent infections

Personal and Family History

A positive family and or personal history is a significant factor

Hemophilia

Hemophilia is the most common hereditary bleeding disorder. It is caused by clotting factor deficiencies

Anemia

Assess type:

Hemolytic
Aplastic
Sickle cell (genetic)
Thalassemia (genetic)
Pernicious
Iron-deficiency
Macrocytic

HIV positive/AIDS

Assess if HIV status is known

Fatigue/Night sweats/Frequent fever/Weight loss

Fatigue -- a frequent subjective complaint in hematolymphatic disorders

Night sweats, fever, weight loss, swollen glands, bleeding, bruising, and fainting are common signs

Transfusions/Toxins/Multiple sexual partners/IV drug use

Transfusions -- Some hematolymphatic disorders are transmitted via blood (HIV); assess date and reason for transfusion

Toxins -- Exposure to occupational radiation or radiation therapy is a risk factor in the development of certain hematolymphatic disorders (leukemia, lymphoma, aplastic anemia, or multiple myeloma)

Multiple sexual partners -- Increased risk of HIV infection and Hepatitis B

IV drug -- Increased risk of HIV infection and Hepatitis B

Hemtolymphatic system testing

Blood tests/ urine testing

Variety of tests may be ordered: Complete blood count (CBC), prothrombin time (PT), partial thromboplastin time (PTT), erythrocyte sedimentation rate (ESR), hematocrit (Hct) and hemoglobin (Hgb), Schilling test, urine urobilinogen, Bence-Jones protein assay and erythrocyte life span testing

Current information

Physician -- List primary physician

Medications -- List all medications; assess for side effects with blood formation or function

HEMATOLYMPHATIC SYSTEM ASSESSMENT FORM

Chief Complaint

Patient's Statement__________Onset________Symptoms__________

Frequency________Duration__________Other areas affected__________

Have you had this before?__________Date __________

What have you done for this?__________

What do you think caused this to happen?__________

What changes have you had to make in your life?__________

Personal and Family History

	Patient Date	Family member		Patient Only Date
Hemophilia (type)	____	____	HIV positive/AIDS	____
Anemia (type)	____	____	Fatigue	____
Immune system problem: Cancer, Lupus	____	____	Night sweats	____
			Frequent fever	____

Weight loss__________Lumps under arm pits, neck or groin__________

Frequent infection__________Types__________

Bleeding from gums, rectum, nose, between periods__________

Easy bruising__________Fainting________Dizziness__________

Blood transfusion history__________

Exposure to radiation/toxic agent__________Type__________

Number of sexual partners__________IV drug use__________

Hematolymphatic System Testing

Blood tests__________Urine tests__________

Lymphangiogram______Biopsy______Bone marrow exam______Bone scan______

Current Treatments/Medications

Who is your physician__________Phone__________

Medications:

Name__________Dose______Frequency______Route________

Name__________Dose______Frequency______Route________

Name__________Dose______Frequency______Route________

HEMATOLYMPHATIC SYSTEM PHYSICAL ASSESSMENT FORM

Inspection:

General appearance______________________________

Shortness of breath or activity intolerance______________________________

Syncope______________Skin color______________Skin temperature______________

Ecchymosis______________________Petechiae______________________

Skin lesions: Location______________Size______________Color______________

Temperature of surrounding skin______________________Drainage______________

Mucous membrane color______________________Gum condition______________

Mouth ulcerations/infections______________________________

Bleeding: Nose, gums, sores, rectum, hematuria______________________________

Ophthalmoscopic exam______________________________

Sclera color______________________Conjunctiva color______________

Palpation:

Temperature______________Route__________Pulse______Resp______B/P____/____

Lymph nodes: head, neck, axillary, epitrochlear, inguinal, popliteal

Location/size/tenderness/______________________________

Palpable spleen______________________Liver______________________

Tender or painful areas______________________________

Assessment Notes

____

____

____

____

Physical Assessment

Physical assessment of the hematolymphatic system is done by inspection and palpation.

The primary task will be to inspect and palpate the lymph nodes for abnormalities. When palpating lymph nodes use a gentle rotating motion. The form on page 160 can be used to complete an assessment of this system.

Patient should be undressed, gowned and covered with a sheet.

Inspection

General appearance

Asses for fatigue, weakness, or cachexia.
Positioning -- What is position at start of assessment?
Does it require excessive effort for position changes?

Shortness of breath/syncope

In hematolymphatic disease there may not be enough RBCs (anemia) to transport oxygen; simple movements may require excessive effort and shortness of breath or even syncope (fainting) may occur

Skin color/ temperature/ condition

Color -- Pallor, cyanosis, or jaundice may be present
Ecchymosis -- Bruising, may be present with low platelets
Petechiae -- Small circular purple discolorations related to abnormality of blood-clotting mechanism
Skin lesions or rashes -- Note location, color and drainage if present
Temperature -- Hot, reddened, dry, edematous skin may indicate infection

Mucous membrane color/ oral condition

Mucous membranes -- Normally moist, pink and intact with no swelling or lesions present
Cyanosis in mucous membranes indicates lack of oxygen
Oral condition -- Assess for ulcers or thrush (often seen in immune system disorders), bleeding gums (low platelets)

Bleeding

Assess for bleeding from any orifice related to clotting factor or platelet disorder
Hematuria -- blood in urine

Ophthalmoscopic examination

May reveal bleeding fundi in coagulation disorders

Sclera and Conjunctiva

Sclera -- Normally white without jaundice or hemorrhage

Conjunctiva -- Normally rose colored

Palpation

Temperature/ Pulse/ Respiration/ Blood pressure

Elevated temperature, rapid pulse may indicate infection
Rapid pulse, wide pulse pressure, heart murmurs and palpitations may indicate the body is compensating for anemia by increasing cardiac output

Lymph nodes

Assess enlarged lymph nodes:
Size -- Enlarged lymph nodes may measure 1 cm to size of tennis ball
Mobility -- Fixed immobile nodes may indicate malignancy
Characteristics -- Hard, soft, spongy, painful, non-tender, red or hot

Spleen/Liver

Spleen in the adult should not be palpable unless enlarged
If spleen and liver are palpated, note location and tenderness

Section III – LABORATORY & DIAGNOSTIC TESTS

The following tables contain the more common laboratory and diagnostic tests used in this system.

Table of Common Laboratory Tests

Test Name	Indications	Comments
Blood urea nitrogen (BUN) 1.7-20.5 micro mol/L Male 10-25 mg/dl Female 8-20 mg/dl	Liver damage	**Regarding Collection:** Use gray-top tube, collect 5 ml Check policy, may require NPO for 8 hours before testing **Results:** Decreased in liver damage or overhydration Increased in kidney damage or dehydration
Bleeding Time 1-6 minutes (Ivy) 1-3 minutes (Duke)	History of bleeding Easy bruising Familial bleeding	**Regarding Procedure:** Cleanse area with alcohol, allow to dry. Ivy - forearm is used with BP cuff Duke - earlobe is used Puncture site 2.5mm deep Time with stopwatch Blot blood drops carefully every 30 seconds until bleeding stops Record time for bleeding to stop in seconds Blot only drops, not incision Assess for aspirin, cold tablet, or alcohol use **Results:** Increased in thrombocytopenia, leukemia, DIC, hemophilia
Blood typing ABO group (Cross matching) Four major blood types A, B, AB, and O	Blood transfusion	**Regarding Collection:** Use red-top tube, collect 10ml **Results:** Patient blood will be assigned to one of the major blood groups and Rh factor will be determined

Complete blood count (CBC) Six components: RBC count Hemoglobin Hematocrit RBC indices WBC count WBC differential	Routine testing Detect anemia Determine blood loss Determine infection Detect blood cell changes	**Regarding Collection:** Use lavender-top tube collect 3ml Keep tourniquet on less than 90 seconds to prevent hemolysis **Results:** See page 168 for the results of CBC testing
Erythrocyte sedimentation rate (ESR)	Acute inflammatory process Infection Cancer	**Regarding Collection:** Use lavender-top tube collect 7 ml Take to lab immediately If left standing sed rate may rise **Results:** Increased in arthritis cancer Decreased in sickle cell anemia, mononucleosis
Hemoglobin (Hgb) Electrophoresis A1 = 95-98% A2 = 1.5-3.5% C = 0% D = 0% F = less than 2% S = 0%	To detect normal hemoglobin types A1, A2, and F To detect abnormal hemoglobin types such as C, M, S	**Regarding Collection:** Use lavender-top tube, collect 10ml **Results:** Hemoglobin F - Thalassemia Hemoglobin C - Hemolytic anemia Hemoglobin S - Sickle cell anemia There are more than 150 types of hemoglobin
Iron 50-150 ug/dl 10-27 umol/L	To assess for iron deficiency/excess due to diet, blood loss or metabolic or problem	**Regarding Collection:** Use red-top tube, collect 5-10 ml Prevent hemolysis of blood sample Note suspected hemolysis on slip Call laboratory for food/drink restrictions **Results:** Decreased - Iron deficiency anemia, cancer, peptic ulcer Elevated - hemolytic and pernicious anemia, thalassemia, lead poison

Partial thromboplastin time (PTT) 60-85 seconds Activated partial thromboplastin time (APTT) 20-35 seconds	To detect clotting deficiencies To monitor heparin therapy Used to determine if adjustments in heparin therapy are needed With heparin APTT may be 1.5-3 times the normal value	**Regarding Collection:** Use blue-top tube, collect 7-10ml Fill tube to capacity Pack sample in ice and transport to lab immediately **Results:** APTT is the most sensitive Decreased in extensive cancer Increased in hemophilia, DIC, leukemia, heparin use, Vit K deficiency
Platelet Count 150,000-400,000mm^3 or 12 0.15-0.4 X 10/L	Bleeding, bruising Leukemia, anemia, chemotherapy	**Regarding Collection:** Use lavender-top tube, collect 5 ml **Results:** Increased in polycythemia vera, trauma, acute blood loss Decreased in some cancers, anemia, liver disease, DIC
Prothrombin time (PT) 11-15 seconds or 70-100 seconds Check laboratory for time	Bleeding, bruising Monitor warfarin or dicumarol therapy Inability to form a clot	**Regarding Collection:** Use black-top or blue-top tube per agency policy, collect 7-10 ml Fill tube to capacity Send to lab immediately **Results:** Decreased in myocardial infarct, pulmonary embolism, or with some medications or vitamin K Increased in liver disease, factor deficiency, leukemia, warfarin or dicumarol therapy
Bone marrow aspiration Normal number of cells with normal size and shape	Hematologic problem leukemia, myeloma, aplastic anemia, Hodgkin's disease	**Pre-Procedure:** Signed consent form is required Explain procedure to patient A local anesthetic will be given Moderate-severe pain at aspiration Position according to site Pre-medication may be given

		Post-Procedure: Apply pressure to site if bleeding Apply sterile dressing or band aid Bedrest for 30 minutes Assess site for bleeding Assess for signs of shock Assess for continued pain (fracture) Post-procedure medication may be given
Bone scan (nuclear scan) No abnormality	To detect bone disease Metastic cancer To assess response to cancer therapy To detect fractures	**Pre-Procedure:** Explain procedure to patient A radionuclide will be given IV After administration, a 2-3 hour waiting period before scan Encourage water drinking (6 cups) Patient needs to void before scan Imaging will take 30-60 minutes Patient should lie still Signed consent may be required All jewelry will not to be removed **Post-Procedure** No activity restrictions
Computerized tomography (CT Scan) With or without iodine contrast dye	To detect tumors To assess enlarged lymph nodes	**Pre-Procedure:** Signed consent form may be needed NPO for 3-4 hours before testing All jewelry and metal is removed Assess for allergy to iodine Patient should lie still May take up to 90 minutes **Post-Procedure:** Assess contrast dye allergy, skin rash, hives, headache, emesis No activity restrictions
Lymphangiography Normal lymph nodes and lymph vessels	Enlarged lymph nodes Hodgkin's disease Metastatic cancer	**Pre-Procedure:** Signed consent form is required Pre-medication may be ordered Contrast dye will be injected over

90 minutes via infusion pump
Patient should lie still
X-rays will be taken post-infusion
X-rays will be retaken in 24 hours
Contrast dye lasts 6-12 months
Assess for allergy to iodine
Post-Procedure:
Skin may be bluish for 24 hours
Urine / feces may be blue for 72 hours
Assess vital signs and incision

The Complete Blood Count (CBC)

The CBC is a blood test with six components; RBCs, WBCs, hemoglobin (Hgb), hematocrit (Hct), RBC indices (MCV, MCH, MCHC), and WBC differential. Platelets are often ordered with a CBC or may be included automatically on the lab slip.

Function/Uses of the CBC:

- Ordered routinely at admission or prior to surgical procedures
- To detect anemias such as aplastic anemia, Cooley's anemia, and iron-deficiency anemia
- To estimate the amount of blood lost during trauma
- To detect blood cell changes
- To detect infections

CBC Values:

	Female	Male	Child	Newborn
RBC $X10^{12}$ /L	4.2-5.4	4.6-6.2	3.8-5.5	3.8-7.2
WBCs $X10^{9}$/L	4.5-11.0	4.5-11.0	6.0-17.00	9.0-30.00
Hgb g/dl	12-16	13.5-18	11-16	14-24
mmol/L	1.86-2.48	2.09-2.79		
Hct	38-47%	40-54%	36-38%	42-54%
volume fraction	0.38-0.47	0.40-0.54		
MCV cu mcg	78-95	78-95	82-91	96-108
MCH pg	27-33	27-33	27-31	32-34
MCHC concentration	32-36%	32-36%	32-36%	32-33%
fraction	0.32-0.36	0.32-0.36	0.32-0.36	0.32-0.33
Platelets cu mm	150,000-400,000		150,000-300,000	
$X 10^{12}$ /L	0.15-0.4	0.15-0.4	0.15-0.3	0.15-0.3

Red Blood Cells (Erythrocytes):

- RBCs make up 99% of blood cells
- Formed in red bone marrow, body produces over 2 million RBCs per second with approximately 35 trillion in circulation
- Life span of a healthy RBC is approximately 120 days
- Each RBC contains hemoglobin which attracts oxygen and carbon dioxide
- RBC count is elevated in persons living at higher altitudes to compensate for a lower atmospheric oxygen concentration

Terms used to describe red blood cells:

Achromatic- colorless, hemoglobin has been dissolved
Erythroblast- immature precursor of erythrocyte
Erythrocytopenia- deficiency of RBCs
Erythrocytosis- increase in circulating RBCs
Polychromatic- uneven staining on laboratory examination
Macrocyte- abnormally large RBCs
Poikilocytosis- abnormal RBC shape

Associated Disorders:

Decreased RBCs	Increased RBCs
Hemorrhage	Polycythemia vera
Aplastic anemia	Dehydration
Cooley's anemia	Increased altitudes
Iron-deficiency anemia	Cor pulmonale
Leukemias	
Multiple myeloma	
Kidney disorders	
Pregnancy	
Excessive hydration	

White Blood Cells (Leukocytes):

- WBCs make up less than 1% of blood cells
- Primary function is to fight infection
- Two major types of WBCs with several subtypes, this is the differentail count
- Granulocytes which are formed in the red bone marrow:
  - Basophils
  - Eosinophils
  - Neutrophils
- Agranulocytes which are formed in the lymph tissue and bone marrow:

Lymphocytes
Monocytes

Terms used to describe white blood cells

Leukopenia -- decrease in WBCs
Leukocytosis -- an increase in WBCs
Leukoblast -- immature precursor of leukocyte
Leukocytopenia -- deficiency of WBCs
Leukocytosis -- increase in circulating leukocytes (infection)

Associated Disorders:

Decreased WBCs	Increased WBCs
Aplastic anemia	Tuberculosis
Pernicious anemia	Pneumonia
Malaria	Meningitis
Alcoholism	Tonsillitis
Antibiotic use	Appendicitis
Some chemotherapeutic	Peritonitis
agents	Pancreatitis
	Gastritis
	Rheumatic fever
	Myocardial infarction
	Burns
	Peptic ulcer
	Leukemias
	Rheumatoid arthritis
	Trauma
	Sickle cell anemia
	Fever
	Convulsions

Hemoglobin:

- Composed of heme (iron-containing pigment) and globin (a protein)
- Primary function is to combine with oxygen to form oxyhemoglobin for transport to cells or to combine with CO2 to form carboxyhemoglobin for transport from the cells to the lungs
- Gives blood its red color
- Elevated in persons living in higher altitudes to compensate for a lower atmospheric oxygen concentration

Terms used to describe hemoglobin

Achromatic -- colorless RBCs due to hemoglobin dissolving
Hemoglobins -- found in sickle-cell disorders
Hemoglobinemia -- excessive hemoglobin in plasma
Hemoglobinopathies -- diseases of hemoglobin

Associated Disorders

Decreased Hgb	Increased Hgb
Anemias	Chronic obstructive pulmonary disease (COPD)
Hemorrhage	Congestive heart failure (CHF)
Kidney disease	Dehydration
Leukemias	Polycythemia
Lymphoma's	Severe burns
Pregnancy	
Slow chronic blood loss	
Thalassemia	

Hematocrit (Hct):

- Measures the percentage of blood volume occupied by blood cells, primarily RBCs
- When drawn by capillary tube Hct may be falsely low
- Hct requires that the blood sample be centrifuged for 2-3 minutes to separate the plasma from the blood cells

Associated Disorders

Decreased Hct	Increased Hct
Anemias	COPD
Hemorrhage	Late dehydration
Kidney disorders	Polycythemia vera
Leukemia's	Transient cerebral ischemia
Lymphomas	
Multiple myeloma	
Peptic ulcer	
Systemic lupus erythematosus	
Vitamin deficiencies	

RBC Indices:

- MCV is the mean cell volume
  Measures the size of RBCs in cubic microns
  Can be calculated: MCV= HCT X 1000 / RBC Count

- MCH is the mean cellular hemoglobin
  Measures the weight of hemoglobin in RBC
  Can be calculated: MCH=Hb X 10 / RBC Count
- MCHC is the mean cellular hemoglobin concentration
  Measures the concentration of hemoglobin in the RBCs sample
  Can be calculated by either formula below
  MCHC= Hgb / Hematocrit or MCHC= (MCH / MCV) X 100

Terms used to describe the red blood cell indices:

Microcytic -- abnormally small RBCs less than 5 microns in diameter or less than 80 femtoliters
Macrocytic -- abnormally large RBCs greater than 10 microns in diameter or greater than 98 femtoliters
Megalocyte -- abnormally large RBC

Associated Disorders

	Decreased	Increased
MCV	Anemia, iron deficiency Anemia, sickle cell Carcinoma Lead poisoning Radiation Thalassemia TRheumatoid arthritis	Anemia, aplastic Anemia, pernicious Hypothyroidism
MCH	Anemia, chlorotic Anemia, iron deficiency Anemia, microcytic	Anemia, blind loop Anemia, macrocytic Anemia, pernicious
MCHC	Anemia, hypochromic Anemia, iron deficiency Excessive hydration Thalassemia	Dehydration, severe

WBC Differential (Diff):

- Provides individual counts of the types of leukocytes to assist in diagnosis
- Values are expressed in percentages

The Granulocytes:

- Neutrophils -- Most common leukocyte (50-70%)

Provide the first line of defense against infections through the inflammatory process by ingesting and killing bacteria and small particles via phagocytosis

- Eosinophils -- 1-3% of leukocytes

  They release chemicals that aid in killing infectious agents
- Basophils -- make up only 0.4-1% of leukocytes

  They release histamine and heparin and other chemicals at sites of injury or infection

The Agranulocytes:

- Lymphocytes make up 25-35% of leukocytes and are divided into two major types:
- B cells that produce antibodies
- T cells that aid in cellular immunity
- Monocytes make up 4-6% of leukocytes

  They are the largest of all leukocytes and ingest large particles via phagocytosis

  Monocytes respond late in illness but live longer than other leukocytes

Associated Disorders

	Decreased	Increased
Neutrophils	Anemia, aplastic Anemia, iron deficieny B_{12} and folate deficiency Dialysis Hypopituitarism Infectious mononucleosis Leukemia Malaria Septicemia, severe Systemic lupus erythematosus Viral illness	Appendicitis Bacterial infections Burns Carcinomas COPD Diabetic ketosis Hodgkin's disease Kidney failure Myocardial infarction Pancreatitis Peritonitis Pneumonia Pregnancy, late-labor Rheumatic fever Rheumatoid arthritis
Eosinophils	Burns Cushing's Syndrome Post-steroid use Trauma	Addison's disease Allergic reactions Asthma Carcinomas COPD Gastritis Hodgkin's lymphoma

		Kidney disease Helminth infection Phlebitis Thrombophlebitis
Basophils	Hyperthyroidism Post-steroid use Pregnancy Rheumatic fever Stress	Chronic myeloid leukemia Hodgkin's lymphoma Polycythemia vera
Lympocytes	Anemia, aplastic Carcinomas, selected Guillain-Barre Kidney disease Multiple sclerosis Myasthenia gravis Systemic lupus erythematosus Trauma	Chickenpox Chronic infections Hepatitis Infectious mononucleosis Lymphocytic leukemias Measles Multiple myeloma Mumps Pertussis Viral illnesses
Monocytes	Anemia, aplastic Lymphocytic leukemia	Hodgkin's lymphoma Infectious mononucleosis Leukemia Multiple myeloma Parasitic infection Recovery period from acute bacterial infection Rheumatoid arthritis Sickle cell anemia Systemic lupus erythematosus Tuberculosis Typhoid Ulcerative colitis

Platelets:

- Primary function is to aid in blood coagulation; after tissue injury platelets stick or adhere to the endothelium and then to each other to form a plug
- Severe decrease leads to hemorrhage
- Formed in bone marrow as fragments of large cells called megakaryocytes

Terms used to describe platelets:

- Thrombocyte -- blood platelet
- Thrombocytopenia -- decrease in platelets

Associated Disorders:

Decreased Platelets	Increased Platelets
Anemia, aplastic	Acute hemorrhage
Burns	Bone fracture
Chemotherapeutic agents	Carcinomas, selected
Pregnancy-induced hypertension (PIH)	Polycythemia vera
Infectious mononucleosis	Postpartum
Leukemias	Trauma
Massive transfusions	
Thrombocytopenic purpura	

Section IV -- PROCEDURES AND CONDITIONS

Acquired Immune Deficiency Syndrome (AIDS)

In 1981, the first known AIDS case in the United States was reported. There is still no cure or vaccine for the HIV virus

General Information about the AIDS virus

- HIV is a retrovirus which contains a single strand of RNA. In the usual cell DNA goes to RNA. Retroviruses contain the enzyme reverse transcriptase that can transcribe the viral RNA in the cytoplasm into DNA, the DNA with the virus can then splice itself to all the DNA cells of the body
- HIV can be lymphotropic (attacks the immune system), making the body too weak to fend off invaders. The bacteria and other infectious agents that the body could normally fight, now become life-threatening
- HIV is very specific when it hits the immune system, attacking the T-cells. T-cells are responsible for cellular immunity and are a vital part of the defense system. Under usual conditions when an antigen (such as a virus, fungus, or parasite) enters the body, a T-cell that has been specifically programmed for that antigen is released into the blood. The T-cells then rush to the site of invasion and attach to the invader, destroying it by releasing cytotoxic substances. T-cells also recruit other leukocytes to help fight the foreign invaders by releasing lymphokines
- HIV prefers T4 lymphocytes (subset of T-cells), also known as T4 helper cells
- When HIV infection is present it invades the T-cells. Once inside a T-cell, it releases its RNA strand which is converted to DNA. The DNA then overtakes the T-cell nucleus where it manufactures more HIV. The infected cell eventually dies and newly-manufactured HIV are released to attack other T-cells. The process continues until the body is depleted of T-cells
- HIV infected lymphocytes are carried in semen and blood
- HIV does not directly cause the infections associated with AIDS. The symptoms are caused by opportunistic infectious agents and cancers that take advantage of the weakened body; this produces the clinical signs and symptoms known as AIDS
- HIV can be neurotrophic (attacks the neurons in the brain or nervous system) and may lead to peripheral neuropathies with

weakness and paresthesia, seizures, hallucinations, and progressive dementia

- AIDS is Acquired Immune Deficiency Syndrome
  Acquired -- Meaning you did not inherit the disease
  Immune -- Referring to the Immune system
  Deficiency -- Meaning the Immune system is weakened
  Syndrome -- A group of signs and symptoms that appear together
- Since HIV can hide in the DNA material of the immune cells, it is not detected by circulating antibodies. In addition, the virus multiplies and mutates very quickly leading to a change in the antigen's appearance. When antigens change quickly or often, the antibodies that are produces in response does not recognize and destroy the virus. These are barriers in discovering a HIV vaccine.

General information about AIDS

- AIDS is a world wide epidemic with millions infected
- AIDS affects male and female, heterosexuals and homosexuals
- AIDS affects children as well as adults
  Approximately 2% of all AIDS patients are under the age of 13. Most of the children affected received the virus from their infected parent prior to birth
- AIDS affects all races
  Approximately 54% of all cases are of the white race
  Approximately 29% of all cases are of the black race, however, they make up only 12% of the population
  Approximately 16% of all cases are of the hispanic race, however, they make up only 6% of the population

Symptoms of AIDS

- Testing positive for HIV does not mean that the individual has AIDS
- Once infected with HIV, one of three things may occur:
  1. The person will remain infected and physically well for a period of months or years
  2. The person will become ill but will not develop an opportunistic infection. The symptoms may be mild, severe, or even fatal. Symptoms include:
    Anorexia
    Blood dyscrasia (anemia, leukopenia, thrombocytopenia)
    Weight loss (greater than 10 pounds)
    Intermittent elevated temperatures
    Night sweats
    Malaise (discomfort, uneasiness)

Skin rashes
Chronic diarrhea
Fatigue and lethargy
Persistent swollen lymph nodes (lymphadenopathy)
Oral thrush (candidiasis)

3. The person will develop AIDS: Any of the symptoms listed above plus one or more of these opportunistic infections or secondary cancers make up the syndrome:

Pneumocystis carinii pneumonia
 Persistent nonproductive cough
 Shortness of breath, dyspnea, tachypnea
Kaposi's sarcoma
 Cancer diagnosed in patients under age 60
 Multiple red, purple plaques or nodes on skin
 Elevated temperature
Non-Hodgkin's lymphoma
 B cell lymphoma or unknown phenotype
 Burkitt on non-Burkitt lymphoma (noncleaved lymphoma)
Primary lymphoma of the brain
 Diagnosed in patients under 60 years old
Chronic cryptosporidiosis
 Disease caused by a protozoa
 Explosive diarrhea
 Diarrhea lasting more than one month
 Abdominal cramps
 Highly infectious
Toxoplasmosis
 Disease caused by a protozoa
 Pneumonitis
 Hepatitis
 Encephalitis
Extraintestinal strongyloidiasis
 Roundworm infection
 Fever
 Severe abdominal pain
 Shock may occur
Isosporiasis
 A parasitic protozoan inhabiting small intestine
 Diarrhea lasting more than one month
Cryptococcoses
 A systemic fungus infection
 May involve any organ of the body
 Brain or meninges often involved
Histoplasmosis
 A systemic fungal respiratory disease
Cytomegalovirus infection
 Herpes virus

Infection of an organ other than the liver, spleen, or lymph nodes in patient over 1 month of age
Cytomegalovirus retinitis (loss of vision)
Candidiasis
In the esophagus, trachea, bronchi, or lungs
Chronic mucocutaneous or disseminated herpes simplex virus
Mucocutaneous ulcer lasting longer than 1 month
Bronchitis, pneumonitis or esophagitis over 1 month
Progressive multifocal leukoencephalopathy
Tuberculosis not normally pathologic in man
Mycobacterium avium (TB of birds) disseminated
Chronic progressive pulmonary disease
Extrapulomonary (found outside lungs as well)
HIV encephalopathy
HIV dementia or AIDS dementia
Anyone who is HIV positive is a carrier and can infect other individuals even if they have no symptoms

How is AIDS transmitted

HIV is not transmitted by casual contact or even close nonsexual contact (social kissing, touching). Transmission requires an exchange of body substances containing cells infected with HIV (blood, blood plasma, semen, vaginal secretions). While the virus has been found in tears and saliva, transmission of HIV has not been reported from these body fluids. The three major routes of transmission are:

- Sexual contact with an infected partner. Sexual contact may vaginal, anal, or oral; heterosexual or homosexual. Infected lymphocytes can be transferred to another individual during sexual contact through minute breaks in skin and mucosa. Groups at particular risk include anyone who has had male-to-male sexual contact since 1977, anyone having sexual contact with a prostitute or who has had more than one sexual partner since 1977
- Contamination with HIV infected blood. The virus is carried in the blood and may be transmitted by contaminated hypodermic needles (intravenous drug abusers, accidental needle sticks in health care workers) where HIV is transferred directly from the blood an infected person into the blood of another person. HIV can be transmitted by transfusion of infected blood. Blood used in transfusion is carefully screened and tested for HIV
- Transplacentally leading to congenital infection. If the mother is infected with HIV before or during the pregnancy, the child is at risk. The risk of an infected woman passing HIV to her child is 15-40 percent. Approximately 80 percent of pregnant women with HIV are asymptomatic. Studies have also shown that HIV

is present in breast milk; although rare cases of transmission via breast milk have been reported.

How can an individual avoid getting AIDS ?

- Avoid promiscuous sex: A monogamous relationship with an uninfected partner is recommended
- Encourage informed sex. Ask potential sexual partner any risk factors and avoid sexual contact with anyone in a risk group
- Use a condom when the sexual history of the partner(s) is unknown or use of intravenous drugs is suspected. Condoms are not fail proof and should be worn throughout intercourse
- Avoid illegal intravenous drugs and sharing needles
- Always wear latex gloves when in contact with blood or other body fluids. If blood should contact skin, wash area immediately

Testing for the AIDS antibody

There are two tests used to check for HIV, ELISA and the Western Blot test. The following should be taken into consideration when testing is done:

- A negative test result does not mean that the person has not been exposed to AIDS: It can take from two weeks to eighteen months for antibodies to appear in the blood. It is believed most individuals will be positive after six months
- A positive ELISA screening test should be repeated, if results are still positive, HIV status should be confirmed with the more specific Western blot test to rule out false-positive result
- The test does not show whether the person has AIDS, it indicates only that they have antibodies to HIV
- A positive HIV without symptoms does mean the person can spread HIV
- It can take 10 years or longer for AIDS to develop after HIV infection

Health Care Workers and AIDS

Steps to avoid HIV exposure on the job:

- Maintain good skin care to prevent dry, cracked skin
- Wear gloves and/or gowns whenever you anticipate contact with urine, feces, saliva, pus, blood or any body secretion
- Wear goggles and a mask if blood or body fluids are likely to splash into eyes or mouth or for coughing patients
- Wear gloves when starting intravenous infusions
- Never recap a needle
- Dispose of all syringes in a designated container

- Report needle sticks and file an incident report
- Wear gloves to clean blood or body fluid spills using a bleach solution (1:10) or other disinfectant approved by the CDC
- Articles contaminated with blood or body secretions should be double-bagged and labeled

Treatment of AIDS

- There is no cure or vaccine
- With zidovudine (AZT), AIDS patients may live several years after confirmation of the HIV infection
- Treatment is aimed at symptomatic problems due to opportunistic infections
- Dietary therapy involves a high-calorie, high-protein diet served in small, frequent meals. Total parenteral nutrition may be required in critically-ill patients
- Medication therapy includes several medications to combat the virus or the opportunistic infections:

  Antiviral agents
  Acyclovir
  Dideoxyinosine (ddI, Videx)
  Ribavarin (Virazole)
  Zidovudine (AZT, Retrovir)
  Antibiotics
  Sulfamethoxazole (Bactrim, Gantanol)
  Trimethoprim (Proloprim)
  Pentamidine (Nebupent, Pentam)
  Interferon
  Interferon (Roferon, Intron)
- Management of respiratory condition:

Assessment of vital signs and pulse oximetry

Oxygen administration as needed

Positioning (semi-Fowler's)

Chest physiotherapy (CPT)

Postural drainage

Incentive spirometry

- Administration of blood products:

Fresh frozen plasma (FFP)

Platelets

Packed red blood cells (PRBC)

- Control of infection through use of universal precautions

BLOOD TRANSFUSIONS

Blood Products and Indications for Use:

Product	Indication	Comments
Whole blood no elements removed	Hemorrhage Open heart surgery Exchange transfusion	450cc/unit approximate
PRBC packed red blood cells	Most often used Chronic/acute loss GI bleeding Surgery, trauma	250-300cc/unit approximate May dilute with saline Should infuse within 4 hrs
Platelets	Low platelet count <20,000-50,0000	50cc/unit Generally ordered in multiple units; may be given IV push
Washed RBCs	WBCs almost completely removed Renal failure Hx of transfusion reactions	300cc/unit approximate Very expansive Chance of reaction less
Cryoprecipitate	Hemophilia A Von Willebrand's fibrinogen deficiency	10cc/unit approximate Contains Factors VII & XIII
Fresh frozen plasma	Undiagnosed bleeding Liver disease >10 transfusions Immune globulin deficiency	150-250cc/unit approximate Thaw about 1 hour Factors II, VII, IX, X,XI, XII, XIII; heat labile to V, and VII
Rho gam	Rh-mom with Rh+ baby to provide antibody to Rh factor	Given <72 hrs of birth Rho D immune globulin
Albumin 5% no need for compatibilit	Acute blood loss	Plasma expander made from plasma precipitate Available in most pharmacies
Albumin 25%	Burns Hypoalbuminemia	Plasma volume expander Draws ECF into circulation

General Guidelines for Transfusing Blood Products:

Check physician's orders for type, amount, and when transfusion is to be given.

- Obtain blood sample for type and crossmatch
- Check patient identification band prior to drawing and sending sample
- Use a large-gauge (18) needle to prevent hemolysis which may interfere with compatibility testing
- When blood is ready for transfusion, prepare the blood tubing.
  A tubing with two connectors is used: one is for the blood product and one for normal saline (NS)
  Flush the line with saline according to the package directions
- Start the infusion of NS at a KVO (keep vein open) rate
- When blood arrives, compare it to the physician's orders
  Check with a second registered nurse to be sure it is the correct blood product, donor number, ABO group and compatibility, correct patient, correct identification numbers, correct time to be administered and expiration date
- Check blood bag for punctures, product color and consistency; return the product if any problem is noted
- At the bedside check the patient's identification band for correct name, hospital number and blood identification numbers If any discrepancies exist, **do not** administer
- Obtain baseline vital signs and record
- Attach labeled bag to blood tubing and close off the NS, allowing blood to begin to flow
- Record date and time transfusion was begun
- Monitor patient closely: Assess vital signs after first five minutes then every 15 minutes X2, and every 30 minutes thereafter
- Blood should be transfused within four hours, check orders for administration time
- Assess for any transfusion reaction to include:

Increase in temperature greater than 1.8 degrees above baseline

Tachycardia
Drop in blood pressure
Onset of back or chest pain
Rash
Sudden unexplained feelings of doom
Wheezing
Cyanosis

- If any adverse reaction occurs:
  STOP THE TRANSFUSION
  Open the clamp to the saline
  Stay with the patient and assess the vital signs
  Have someone call the physician and blood bank
  Recheck the blood component against the identification band
  With severe symptoms, the transfusion will not be resumed
  Blood component and tubing are sent to the lab and then the blood bank
  Blood and urine samples are obtained
  Physician orders treatment based on patient's condition
  For allergic transfusion reactions (hives or rashes), an antihistamine may be ordered and the transfusion continued

Rules for Transfusion of Packed Red Blood Cells (PRBC'S)

- Never add any medications to blood components
  Hemolysis may occur
- Use only normal saline (NS) when transfusing blood
  D5W may lyse the red blood cells being transfused
  Lactated Ringers may cause clotting
- Administer blood only after double checking with another nurse
  Protects against administration of wrong blood type
- Never store blood in the unit refrigerator
  If transfusion is delayed return blood to the blood bank STAT
- Always use a peripheral line started with a large gauge needle
  A central line is only used if ordered as use may lead to dysrhythmias
- For multiple transfusions, change filter every four hours to reduce the risk of bacterial growth

Charting Tips

Chart the following:

Reason for transfusion
Total amount to be transfused, number of this unit
Donor Number, ABO type, Rh type, expiration date of unit
Second person who checked blood component and title
Date and time begun
Vital signs at onset, after 5 min, 20 min, 30 min
Vital signs at the end of transfusion
Any reaction noted and actions taken
Time and date of transfusion is complete

Interventions for Patients with Hematolymphathic Problems

When healthy organs or cells are destroyed due to disease (AIDS or cancer) or chemotherapeutic agents, the following conditions may occur:

Anemia -- reduction in RBCs
Leukopenia -- reduction in WBCs
Thrombocytopenia -- reduction in platelets
Pancytopenia -- reduction in all blood cells

Interventions are listed below:

Anemia

- Monitor RBC level
- Monitor hematocrit and hemoglobin levels
- Assess for fatigue, dizziness, chills or shortness of breath, all of which are symptoms of a low red blood cell count
- Instruct patient on chemotherapy to report any symptoms
- Encourage periodic rest periods during the day
- Encourage foods high in iron such as green leafy vegetables, liver, and red meats
- Encourage getting out of bed slowly to prevent or reduce dizziness
- Assess the need for oxygen therapy and initiate if indicated
- Administer blood transfusions as ordered

Leukopenia

- Monitor white blood count and differential
- Inform physician of any abnormal values
- Assess for elevated temperature above 100 degrees F which may indicate infection
- Assess for chills, sweating, diarrhea, burning on urination, sore throat or coughs
- Assess mouth for redness, open sores or ulcers, known as stomatitis or oral mucositis
- Instruct patient to report any symptoms
- Instruct compromised patient how to avoid infection:
  Wash hands frequently, especially before eating and after using the restroom
  Avoiding large crowds; avoiding anyone known to have respiratory or other contagious illnesses
  Thorough cleansing of any cuts, scrapes, or wounds at once with soap and water

To use an electric shaver, to avoid abrasions from a razor
No cutting or tearing cuticles; use a commercial cuticle softener
Avoiding squeezing or scratching pimples or skin blemishes
Cleaning the anal area thoroughly after each bowel movement
Postponing elective dental work
Postponing elective surgical procedures

- If stomatitis is present instruct the patient to rinse mouth before and after eating with a normal saline solution: 1 tsp. salt in 1 quart water

Thrombocytopenia:

- Monitor platelet count
- Assess for bruises and petechiae
- Assess for bleeding from any orifice to include anus, gums or in stool, urine, saliva, phlegm, or sputum
- Apply pressure on any venipuncture sites for 5-10 minutes when low platelet count is present
- Avoid unnecessary venipunctures or intramuscular injections
- Monitor number of peri-pads during menstruation to assess for unusual bleeding
- Instruct patient in how to avoid bleeding:

  Use of an electric shaver
  Avoidance of any over-the-counter medication containing aspirin
  Wearing shoes at all times, to prevent trauma to feet

- Instruct patient to avoid using dental floss; use a soft foam toothbrush
- Avoid hot or spicy foods, abrasive foods, acidic foods or liquids such as citrus juices
- Encourage eating soft foods when ulcers are present

Pancytopenia

All of the above interventions are appropriate for someone with pancytopenia

The table below contains common problems and interventions that may be helpful when caring for a patient with problems in the hematolymphatic system

Problem	Intervention
Bad breath (Halitosis)	Offer mints and gum Have normal saline available to use as a mouth rinse

Loss of appetite	Provide small, frequent meals Provide attractive meals Offer high-calorie, high-protein diet Encourage rest prior to meal
Nausea/vomiting	Offer antiemetics as needed Remove bedpans immediately after use Serve foods at room temperature, these are better tolerated than hot or cold foods Do not serve liquids with meals or within 1 hour of eating Avoid foods that are fried, fatty or sweet, as these increase nause Avoid foods with strong odors Instruct patient to eat slow and chew the food well to promote digestion Provide dry foods like toast, dry cereal, or crackers Provide cool unsweetened drinks between meals (flat soda, ginger ale, or apple juice) Discuss relaxation techniques, music therapy or other non-traditional approaches
Dry mouth	Provide ice chips Increase fluid intake Offer sugarless gum or hard candy Moisten dry foods with gravy or sauce prior to serving Offer lip balm or petroleus jelly for dry lips
Stomatitis	Assess condition of sores or ulcers daily Avoid serving highly spiced or acid foods Provide diet of soft unseasoned foods Discourage smoking Avoid commercial mouth washes, use saline instead as these may irritate the tissues Brush teeth 3-4 times daily Offer antacids, Milk of Magnesia Offer warm salt water, or hydrogen peroxide to rinse mouth
Diarrhea	Increase fluids/force fluids; include water, weak tea, broth, apple juice For severe diarrhea provide a clear liquid diet Provide a low-residue diet Monitor input and output Avoid serving oily or greasy foods Avoid caffeine

	Avoid milk and milk products Encourage meticulous anal care after each bowel movement Provide anti-diarrheal agents as ordered
Constipation	Provide high-fiber diet Increase fluid intake Increase physical activity Provide stool softener or laxative as ordered
Hair loss (Alopecia)	Hair loss is usually temporary Encourage use of scarves, wigs, hats prior to severe hair loss
Sexual dysfunction	Assess for loss of sexual function Encourage open discussion of fears with partner and physician Explain that fatigue, which often accompanies cancer treatment, can affect sexual desire Encourage rest periods if fatigue is a problem Cessation of menstruation or irregular menstrual cycles can occur with chemotherapy treatment Assess for dryness, itching of vaginal tissue, provide topical medication as ordered Encourage the use of birth control medications during chemotherapy to prevent pregnancy

Section V-- DIETS

Diets Used In Hematolymphatic Disease

The following dietary information applies to the hematopoietic system, foods high in iron are included for their importance in the synthesis of hemoglobin.

Iron-rich foods

Indications: Iron-deficiency anemia, pregnancy, infancy, and childhood; increase need in presence of slow blood loss; diseases that interfere with iron absorption.

Comments: Iron is needed for the synthesis of hemoglobin. The average American diet has approximately 6 mg of iron per 1,000 calories. The Recommended Dietary Allowance for iron is:

Infants	6-10 mg
Children	10 mg
Adolescent Males	12 mg
Adolescent Females	15 mg
Adult Males	10 mg
Adult Females	18 mg
Postmenopausal Females	10mg
Pregnant Females	*30 mg; supplement recommended
Lactating Females	*15 mg; supplement recommended

*The recommended allowance of iron is high in the pregnant female, and it would be difficult to plan menus to meet iron needs.

The majority of dietary iron is supplied by the meat group (approximately 40%), followed by the bread and cereal group.

Restricted Foods: No foods are restricted

Allowable Foods: All the Basic Four foods should be included for a balanced diet, the following are high in iron:

Meats
- Pork Liver -- 140 mg per serving
- Beef Liver -- 70 mg per serving
- Beef Kidney -- 60 mg per serving
- Dried Beans -- 30 mg per cup
- Beef -- 30 mg per cup
- Pork -- 30 mg per cup

Breads
- Farina cereal -- 90 mg per cup

Fruits
- Prune juice -- 60 mg per cup
- Dates -- 30 mg per cup
- Raisins -- 30 mg per cup

Vegetables
- Spinach -- 30 mg per cup

Other Interventions: Supplements are recommended during pregnancy. During pregnancy maternal blood volume increases, requiring more hemoglobin production and thus more iron. The pregnant women also supplies her unborn child with iron for the first 3-6 months of life. Ascorbic acid aids in the absorption of iron and should be encouraged

High-protein, high-calorie diet

Indications: Illness, infection, burns, cancers

Comments: Anorexia, altered taste sensations, nausea and vomiting are problems during the treatment and recovery periods from cancer. A high-calorie, high-protein diet is recommended

Restricted Foods: No foods are restricted; however, food associated with nausea are:
- Rich, sweet foods
- Greasy, fried foods
- Liquids at mealtime
- Foods with strong odors

Allowable Foods: All foods are allowed, unless they are not tolerated by the individual. The following foods are high in protein/calories and can be easily added to the diet:

To add protein:

During food preparation

Add 2 tbs. powdered milk to sauces, gravies, soups, hot cereals, ground meats, eggs, casseroles

Substitute milk for water in recipes for cookies, soups, puddings and cocoa

Add meats to noodles, soups, sauces, potatoes, rice, casseroles

Add cheese to noodles, vegetables, soups, sauces, sprinkle on potatoes, rice, breads

Put cottage cheese in spaghetti sauce, meat sauces, bread or use as vegetable dip

Use peanut butter for vegetable dip, spread on cookies, brownies, pancakes, waffles, cake, vegetables (celery), fruit (apples, bananas, pears) or bread
Add nuts to breads, cookies, cakes, ice cream, cheese or use as a snack

Adding calories:
Add butter to soups, vegetables, potatoes, rice, hot cereals, pancakes, waffles, breads, casseroles
Add mayonnaise or salad dressing to sauces, eggs, sandwiches, for vegetable dip or for salads
Spread honey on toast, use it in coffee or teas
Add yogurt or sour cream to potatoes, vegetables, chili, gravies, or dressings and dips
Put whip cream on hot cocoa, pancakes, pudding, jello, pie or fruit
Bread meats and vegetables

Comments: Meals should be served frequently, four to six small meals may work better than three large meals

Weight should be taken daily at the same time, using the same scale to assess for effectiveness of dietary interventions

Section VI -- DRUGS

The tables below supply only general information: a drug handbook should be consulted prior to administering any unfamiliar drug.

ANTIVIRAL AGENTS

Action: To inhibit viral activity and replication.

Indications: Some medications are used for HIV prophylaxis or to treat AIDS.

General Comments: The four antiviral medications listed here are being used in the treatment of AIDS.

Examples of Drugs in this Classification:

Generic	Trade	Comments
Acyclovir	Zovirax	Used in genital herpes, herpes zoster, and AIDS Oral, IV, and topical forms Acyclovir sodium is for IV only IV infusion over one hour Monitor I&O with IV use
Dideoxyinosine	ddI, Videx	Used in advanced HIV infection in patients who cannot tolerate AZT Not toxic to bone marrow Taken twice daily Assess for peripheral neuropathy Assess for S&S of pancreatitis Less costly than AZT; $2,000/year
Ribavirin	Virazole	Aerosol treatment only Used in RSV viral infection, AIDS and HIV, and herpes Monitor respiratory function and fluid status during therapy Keep accurate I&O records
Zidovudine	AZT, Retrovir	Used in HIV infection Oral medication Important to take every 4 hours Baseline blood counts are needed Anemia and granulocytopenia may occur from bone marrow toxicity

Use of drug does not reduce transmission risk of HIV
Costly medication: $10,000/year

FOLIC ACID

Action: Needed for the formation of red blood cells

Indications: Folate deficiency, anemias, alcoholism, liver disease, increased need in pregnancy

General Comments: Oral contraceptives have been linked with folate deficiency, as have alcohol, barbiturates, and some anticonvulsant

Examples of Drugs in this Classification

Generic	Trade	Comments
Folic Acid	Folacin Folvite Novofolacid	History of drug and alcohol use should be taken Dietary recall should be obtained

HEPARIN ANTAGONIST

Action: Antidote for heparin overdose

Indications: Heparin overdose

General Comments: Monitor blood pressure and pulse closely during use, and until stable for 3 hours. Coagulation studies need to be done periodically during treatment

IRON PRODUCTS

Action: Iron is required for the synthesis of hemoglobin. Indications: Iron-deficiency anemia, increased need during pregnancy, infancy and childhood, slow blood loss

General Comments: Encourage a diet high in iron. Administer in between meals with liquid. Do not take with any antacids or milk products. Iron products may cause discoloration of the stools to a dark green or black

Examples of Drugs in this Classification

Generic	Trade	Comment
Ferrous Sulfate	Feosol Fer-in-Sol Fer-iron	Given PO Give liquid preparations through a straw, as solution may discolor teeth Foods high in ascorbic acid increase iron absorption
Iron Dextran	Feostat Feronim Hematran Imferon K-Feron	Given IM , IV Administer 1st dose cautiously and assess for allergic reaction IM give only in UOQ of buttock Use Z track administration technique

VITAMIN K

Action: Needed for formation of blood clotting Factors II, VII, IX, X

Indications: Disorders interfering with Vitamin K manufacture and absorption to include ulcerative colitis, intestinal resections, enteritis, obstructive jaundice, as well as for antidote for anticoagulant overdose

General Comments: Dosage dependent upon Prothrombin time and presence of bleeding tendencies, a deficiency of this vitamin in a healthy individual is not likely.

Examples of Drugs in this Classification

Generic	Trade	Comments
Menadione	Vitamin K_3 Menaphthone	Given PO Check with physician regarding dietary supplement(s)
Menadiol	Vitamin K_4	Given PO, SC, IM or IV
Phytonadione	Vitamin K1 AquaMEPHYTON Mephyton Phylloquinone	May be given PO, SC, IM, or IV Not an antidote for Heparin Antidote for Coumadin Given to neonates for prophylaxis; IV use for emergency only Check container for route as some trade names are not for IV use

Section G -- GLOSSARY

ABO	Blood type groups
Activated partial thromboplastin time (APPT)	Laboratory test to detect clotting deficiencies or a clotting ability of the blood
AIDS	Acquired Immune Deficiency Syndrome, caused by HIV
Agranular leukocytes	Lymphocytes and monocytes
Alopecia	Loss of hair
Anemia	Decrease in functional RBCs or hemoglobin
Aplastic anemia	Anemia from bone marrow disease or destruction
Basophils	Type of WBC, known to release heparin, histamine and serotonin
Blast	Immature cell
Cachexia	Malnourished condition with muscle wasting related to severe or prolonged illness
Carcinoma	Pertaining to cancerous growth
Differential count	Calculation of the percentage of each type WBC
Disseminated intravascular coagulation (DIC)	Blood coagulation is altered, leading to potential for massive bleeding
Dyspnea	Difficult breathing with noted increase in work above normal respirations
Ecchymosis	A bruise, an irregular shaped area on the skin that is blue-black, green to yellow, or brown caused a hemorrhagic spot

ECF	Extracellular fluid
Embolus	Any foreign material in circulating blood
Eosinophils	Type of WBC, usually increased in allergic reactions
Epistaxis	Nosebleed
Erythrocyte	Red blood cell (RBC)
Erythropoiesis	Formation of RBCs
Extravasation	The escape of fluids and/or medications into the surrounding tissue
Erythropoietin	Hormone that stimulates erythropoiesis
Fractionated blood	Blood separated into its components
Granular leukocytes	Neutrophils, eosinophils and basophils
Hematocrit (Hct)	Value of the blood cell volume in whole blood, primarily RBCs
Hematoemesis	Vomiting with blood
Hematoma	Accumulation of blood in tissue, organ or space
Hematuria	Blood in urine
Hemoglobin (Hgb)	Red pigment that carries iron in RBCs
Hemolysis	Rupture of RBC with loss of hemoglobin
Hemophilia	Hereditary disease due to a deficiency in clotting factors in the blood
Hemoptysis	Spitting up blood
HIV	Human immunodeficiency virus, the virus that causes AIDS
Hypoxia	Cellular oxygen deficiency

Leukemia	Uncontrolled, unorderly multiplication of WBCs; may be acute or chronic
Leukocyte	White blood cell (WBC)
Leukopenia	Decrease in number of WBCs in blood
Lymphoblast	Precursor for lymphocyte
Lymphoma	Malignant growth of tissue in the lymphatic system
Megakaryoblast	Precursor for platelets
Melena	Black, tarry stools
Monocytes	WBCs generally increased in chronic infections
Monoblast	Precursor for monocyte
Myeloblast	Precursor for neutrophils, eosinophils and basophils
Myelosuppression	Bone marrow suppression
Neutrophils	Most active WBC during early bacterial infections, destroy bacteria by phagocytosis
Pancytopenia	Decrease of all elements in the blood (RBC, WBC, platelets)
Partial thromboplastin time (PTT)	A laboratory test used to detect clotting deficiencies
Perniciousanemia	Anemia produced by inadequate intrinsic factor
Petechiae	Small pinpoint hemorrhages in skin or mucous membranes
Polycythemia	An excess of red blood cells
PTT	Partial thromboplastin time, a laboratory test used to detect clotting deficiencies

Septicemia	Disease resulting from toxins or bacteria in the blood
Serum	Plasma minus clotting proteins
Splenomegaly	Enlargement of the spleen
Syncope	Fainting, usually due to an inadequate blood flow to the brain
Thalassemia	Hereditary anemias in which hemoglobin production is abnormal, more common in Mediterranean and Southeast Asian populations
Thrombus	A clot lodged in a vessel or heart cavity
Thrombocyte	Platelet
Thrombocytopenia	Decrease in number of platelets in blood
Type and hold	Blood sample is typed and screened for irregular antibodies but blood is not matched to specific donor units
Type and crossmatch	Blood sample is typed and the patient's blood is matched to specific donor units

Gastrointestinal System

GASTROINTESTINAL SYSTEM

Table of contents

Section I -- OVERVIEW

Components

This system contains the teeth, tongue, salivary glands, liver, gallbladder, pancreas and appendix which are accessory organs to the gastrointestinal (GI) tract. The GI tract is composed of the pharynx, esophagus, stomach, small intestine (duodenum, jejunum, and ileum), appendix, large intestine (ascending, transverse and descending colon), rectum and anus.

Primary Functions

- Ingestion, digestion and absorption of nutrients for the distribution to body cells via the circulatory and lymphatic systems
- Elimination of indigestible products and waste from the body
- Assistance in maintaining the balance of fluids within the body
- Variety of vital metabolic functions performed by the liver:
  Production of anticoagulants (heparin)
  Production of prothrombin, fibrinogen and albumin
  Destruction of RBCs
  Storage of vitamins (A, D, E, K) and minerals (copper, iron)
  Manufacture of bile (800-1000 ml/day)

Physiology

Ingestion of food or eating is necessary to maintain life. The cells in the body need the nutrients food provides. These nutrients are released from the food we eat; this is done by our digestive tract. The process of digestion starts when we see or smell food, the salivary glands begin to secrete saliva in anticipation of eating. The three pairs of salivary glands inside the mouth are the parotid gland, sublingual gland and the submandibular gland. The parotid gland secretes saliva which contains the enzyme amylase; the sublingual glands secrete amylase mixed with mucous and the submandibular glands secrete mostly mucous. The salivary amylase begins to break down polysaccharides (starches and sugars) while the food is still in the mouth. The mucous helps lubricate the food.

Once the food enters the mouth, the teeth chew the food (mastication), breaking it into smaller pieces. The salivary glands continue to secrete, and the food is formed into a mass called a bolus. Once the bolus is formed, it is moved towards the back of the mouth by the tongue where it enters the oropharynx and is swallowed (deglutition). During swallowing, the epiglottis covers the larynx so food does not enter the respiratory system. The food travels to the esophagus where it is transported to the stomach via peristalsis.

The Digestive System

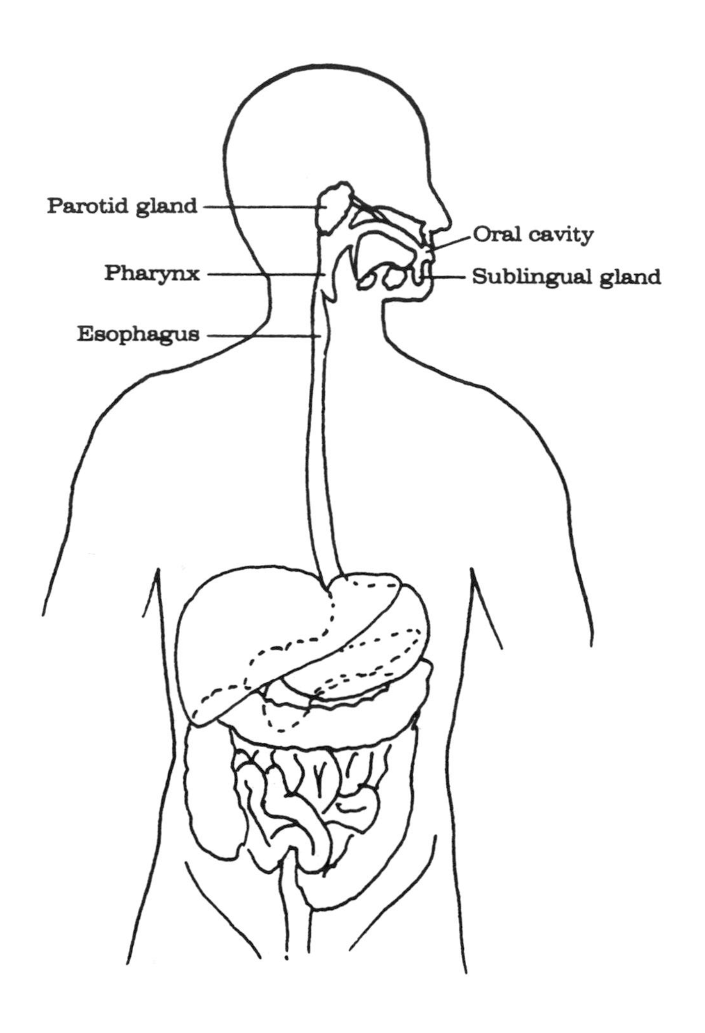

Figure 5A

The Digestive System

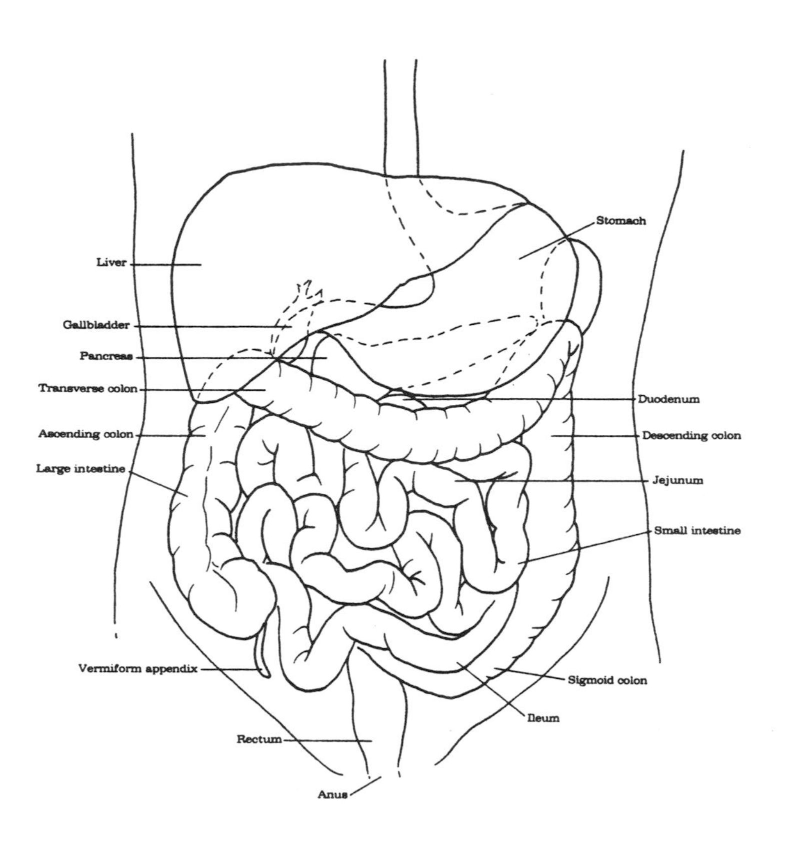

Figure 5B

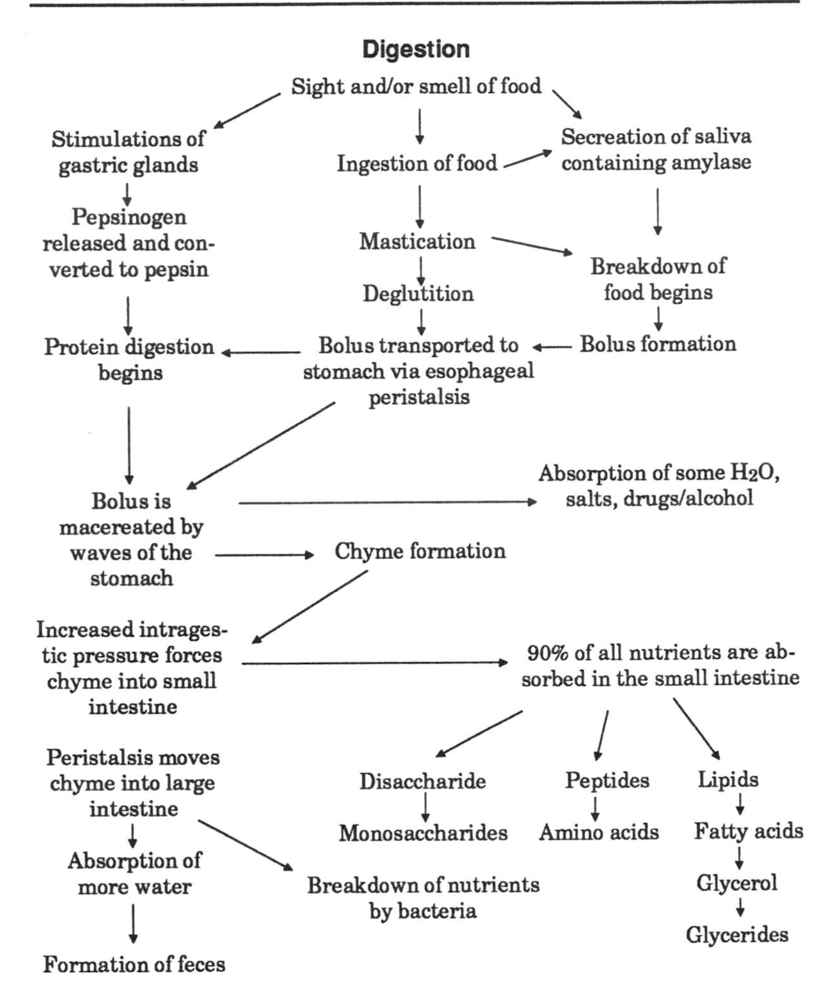
Digestion
Sight and/or smell of food
Stimulations of gastric glands
Ingestion of food
Secreation of saliva containing amylase
Pepsinogen released and converted to pepsin
Mastication
Breakdown of food begins
Deglutition
Protein digestion begins
Bolus transported to stomach via esophageal peristalsis
Bolus formation
Absorption of some H_2O, salts, drugs/alcohol
Bolus is macereated by waves of the stomach
Chyme formation
Increased intragestic pressure forces chyme into small intestine
90% of all nutrients are absorbed in the small intestine
Peristalsis moves chyme into large intestine
Disaccharide
Peptides
Lipids
Monosaccharides
Amino acids
Fatty acids
Absorption of more water
Breakdown of nutrients by bacteria
Glycerol
Glycerides
Formation of feces

Figure 5C

Peristalsis is the successive contraction of the muscles surrounding the GI tract.

Digestion continues in the stomach, the bolus is churned and mixed by gentle peristaltic waves and gastric juice is added. Gastric juice contains pepsinogen which breaks down proteins into peptides after it has been converted to pepsin.

Hydrochloric acid activates the pepsinogen secreted by the parietal cells. As the bolus is broken down in the stomach, it turns to chyme which is a thinner, liquid substance. The chyme is formed into the duodenum in small amounts with each peristaltic wave. The pyloric valve prevents the chyme from returning to the stomach.

In the small intestine digestion continues, enzymes are added from the pancreas (pancreatic juice) to intestinal juice from the cypts of Lieberkühn, and bile is added from the liver to aid in the absorption of fats. Once the chyme reaches the small intestine, enzymes released by epithelial cells on the surface of epithelial cells also aid in digestion. These enzymes, plus the secretions of the pancreas and liver, further break the chyme down so it can be absorbed. The small intestine is able to absorb nutrients by diffusion and active transport through the epithelium; the nutrients then enter the blood capillaries and are carried to the cells by the blood. Peristalsis continues and the chyme moves with the rhythmic waves of muscle contraction down the small intestine. The process of digestion and absorption of nutrients is continuous along the small intestine.

The large intestine is the last part of the GI system. It is here that the remaining products from the food will be processed. The large intestine will absorb more water from the chyme which will now become feces. The large intestine does not secrete enzymes; however, the bacteria located there do continue to break down proteins and carbohydrates, and these are absorbed by the intestinal mucous. The bacteria also synthesize Vitamins B and K which are absorbed. The large intestine also has peristaltic movements. However, they are slower than in other segments of the GI tract. About 3-4 times daily, strong waves of peristalsis occur known as mass peristalsis. These movements push the contents toward the rectum for removal from the body (defecation).

Gastrointestinal Short Facts

- The tongue contains four taste zones:
  Sweet - at the tip of the tongue   Bitter - at the back of the tongue
  Sour - at the posterior sides   Salty - at the anterior sides
- The small intestine is about 21 feet in length.
- The large intestine is about 5 feet in length.

Section II -- ASSESSMENT

Health history tips

- For diabetes, note type:
  Type I -- Insulin-dependent diabetes mellitus (IDDM)
  Type II -- Non-insulin-dependent diabetes mellitus (NIDDM)
- For hepatitis, note the type if known (A, B, C, non-A or non-B) and severity of illness.

Physical examination tips

Request patient to empty bladder prior to examination

- When performing the examination the patient should be supine with abdomen exposed from xiphoid process to groin with genitalia covered
- Avoid touching the patient with cold hands. Rub them together first to warm. Warm the stethoscope prior to use
- Always obtain weight on admission and at least every other day
- On surgical patients record the number, location of any loose or broken teeth prior to intubation
- Pink or purple striae may be present in Cushing's syndrome
- Pulsations from the abdominal aorta may be present in thin persons
- A large, distended abdomen may indicate ascites (fluid-filled)
- Always auscultate prior to palpation or percussion to prevent alteration of bowel sounds
- Use diaphragm for auscultation of bowel sounds
- Note location, size, shape, consistency and tenderness of masses
- Rebound tenderness is pain noted when you withdraw your hands
- Percussion should be done in all four quadrants
- The liver boundaries can be roughly assessed by percussing for dullness (the sound heard over the liver)
- Percussion over the stomach is tympanic due to the gastric air bubble
- Kidneys are normally difficult to palpate. An enlarged, palpable kidney should be reported

ASSESSMENT OF THE GASTROINTESTINAL SYSTEM

Patient's Statement________________Onset__________________Symptoms________

Frequency______________Duration__________________Other areas affected__________

Have you had this before?_________________ What treatment was given?___________

Bywhom_________________________Where?_____________________________________

Have you or any one in your family ever had the following

	Patient Date	Family member		Patient Date	Family member
Diabetes	_____	_____	Hepatitis	_____	_____
Ulcers	_____	_____	Bowel disease	_____	_____

Have you ever had the following surgeries (date and reason)

Gallbladder?_______Stomach?_________________Intestines?_________Liver?_______

Or surgery for hemorrhoids?________________________Hernias?___________________

Have you had blood in your stools/vomit?______When?__________How often?_______

Current History

Who is your physician?____

Are you on a special diet?__________What kind?_________Why?__________________

Medication

Name_________________Dose_________________Frequency__________Route________

How often do you have a bowel movement?____________________/___________________

Do you think you have constipation?______________________Diarrhea?______________

Do yo do anything to help with elimination? (drug, food, exercise)__________________

Do you have: Nausea?___________Vomiting?_____________Loss of appetite?________

Excessive belching?_________________Gas?__________Black tarry stools?__________

What is your usual weight?_________Recent loss/gain?__________How much?_______

What did you have yesterday for:

Breakfast	Lunch	Dinner
____________	____________	____________
____________	____________	____________
____________	____________	____________

Was this a typical day for you?____

Exercise?____

Alcohol use?_____________________________Tobacco use?_______________________

GASTROINTESTINAL SYSTEM PHYSICAL ASSESSMENT FORM

Inspection:

Height_______________Weight____________________Frame size________________

Skin color_______________Texture____________ Rashes________ Jaundice ________

Teeth: Cavities ____________ Broken teeth___________ Dentures

Gums: Redness ___________Swelling_______Mouth: Ulcers_______Vesicles_________

Abdomen: Flat_______Rounded________Obese________Striae_____Pulsation______

Symmetrical__________________Visible masses_________Peristalsis______________

Anus: Hemorrhoids___________________________Fissure_________________________

Stool: Brown____Gray_____Bloody_____Tarry_____Chalky_____Dry_____Watery_____

Mucous_________Undigested food_________Foul odor_________Frequency____/____

Emesis: Color_______Blood_________Bile________Undigested food_____Amount_____

Gastric tubes: NG__________________GT____________Drainage________________

Auscultation:

Bowel sounds: RUQ_____LUQ____LLQ_____RLQ_____Borborygmi____Bruits_____

Palpation:

Skin turgor_____________________________Temperature________________________

Liver__________Spleen_____________Kidney: Rt.______________Lt.____________

Masses_________Location________________Size_________Tenderness_____________

Rebound tenderness_______________________Anal sphincter tone_________________

Percussion:

Liver Stomach Spleen

Assessment notes:

____

____

____

____

Laboratory and Diagnostic Tests

The following table contains common laboratory tests used in this system. Values are in SI units.

Name of test **Normal Value**	**Indications**	**Comments**
Ammonia Check agency for method used diffusion 11.1-67.0 micro mol/L Enzymatic 22.2-44.3 micro mol/L Resin 6.7-26.6 micro moL	Severe liver disease Cirrhosis Necrosis Hepatic failure Erythroblastosis fetalis	Blood sample required By-product of protein metabolism Sample should be placed on ice and taken to lab ASAP
Amylase 111-296 U/L	Pancreatitis Perforated peptic ulcer Cholecystitis	Blood sample is required NPO for 1-2 hrs. prior to test Amylase is an enzyme needed to convert starch to sugar
Bilirubin total 1.7-20.5 micro mol/L Bilirubin direct 1.7-5.1 micro mol/L	Liver function study to determine if bilirubin is conjugated and excreted in the bile Measures conjugated bilirubin	Blood sample is required Check agency policy for dietary/medication restrictions Bilirubin is the by-product of hemoglobin breakdown Elevated bilirubin levels occur with damaged or impaired liver function
Electrolytes	To detect imbalances in potassium/sodium	Blood sample is required
Potassium 3.5-5.0 mmol/L (K^+)	Decreased level in: Vomiting, diarrhea, Gastric suctioning, Diabetic acidosis, Trauma, stress, Surgery, burns, Renal disorders Some medications: K+ wasting Diuretics Steroids Antibiotics Increased level in: Acute renal failure Addison's disease	S/S of hypokalemia are: Hypotension Cardiac arrhythmias Muscle weakness Nausea, vomiting Diarrhea Lasix Cortisone Amphotericin, gentamicin S/S of hyperkalemia: Bradycardia Abdominal cramping

	Metabolic acidosis Crushing injuries Some medications: Antibiotics Marijuana Heparin	Anuria or oliguria Tingling or numbness in the extremities Penicillin G potassium
Sodium 136-142 mmol/L (Na^+)	Decreased level in: Vomiting, diarrhea, Burns, trauma, surgery, renal disease, gastric suctioning, diuretics Increased level in: Dehydration CHF, Cushing's Disease, High Na+ diet Some medications: Antitussives Antibiotics Laxatives	S/S of hyponatremia: Hypotension Tachycardia (Muscular weakness Muscular twitching Headaches Lasix S/S of hypernatremia: Thirst Dry, rough skin Dry, mucous membranes Restlessness Tachycardia High in Na^+, or leads to Na^+ retention
Lipase 14-280 U/L	Pancreatitis	Blood sample is required Lipase is an enzyme aiding in the digestion of fats
Occult blood, fecal test negative	GI tract bleeding	Stool specimen is required "Guiac" is a common test performed
Ova & parasites negative	Parasites suspected in GI tract	Collected over a 3 day period to provide maximum opportunity to detect the presence of organisms
Protein 60-78 g/L	Cirrhosis of liver: Cancer of liver Biliary obstruction Chronic liver disease	Blood sample is required Indirect measure of albumin and globulin
Albumin 32-45 g/L	Same as above	
Globulin 23-35 g/L	Same as above	

This table provides information on diagnostic tests.

Name of test	**Indications**	**Comments**
Barium enema	Ulcerative colitis Diverticulitis	X-ray examination using barium contrast via the rectum

	Intestinal polyps Colon cancer	NPO after midnight Laxatives or enemas are given to clean out the intestines prior to test Test takes about 1 hr. Post-procedure: A laxative or enema is given to remove barium from colon
Barium swallow	Esophageal strictures Esophageal tumors Ulcers Hiatal hernias	X-ray examination using barium as a contrast Test takes about 30 min. to 1 hr. Post-procedure: A laxative may be given to eliminate barium from GI tract
Cholecystography	Cholelithiasis Cholecystitis	NPO for 8-12 hrs. before test Assess for dye allergies Contrast dye tablets are taken 12-14 hrs. prior to test; X-Rays are taken High fat drink may be administered and X-rays repeated Takes from 1-3 hrs.
Duodenoscopy Esophagogastroduoden-oscopy Esophgogastroscopy Gastroscopy	Hiatal hernia Esophageal varices Gastritis Gastric ulcers Duodenal ulcers Carcinoma of upper GI tract	Consent form is needed Endoscopic examination NPO 8-12 hrs. prior to testing Premedications are given (sedative/analgesic) Test takes about 1 hr. Post-procedure: No liquids until gag reflex is present Frequent VS
Proctosigmoidoscopy	Intestinal polyps Intestinal fistulas	A consent form is required Heavy meals are avoided prior to testing, check agency policy for restrictions Laxatives or enemas are given prior to test (usually in the morning) A knee-chest or Sim's position will be used Test takes 15-30 min. Post-procedure: Frequent VS until stable
Ultrasonography	Locate abnormalities in GI tract or in the accessory organs	Non-invasive

Section III -- PROCEDURES & CONDITIONS

Total Parenteral Nutrition

Definition

Total parenteral nutrition (TPN) provides all the caloric needs, via the intravenous route, for patients who cannot or will not ingest enough food. TPN provides protein, carbohydrates, lipids, electrolytes and trace elements to prevent malnutrition. The exact contents of TPN is always prescribed by the physician

Considerations When Giving TPN

Prior to TPN, assess:

- The current nutritional status of patient
- Baseline vital signs (pulse, respirations, temperature, BP)
- Patient's admission weight
- Patient's current weight
- Patient's desired weight
- Caloric needs of the patient: (see caloric calculations)
- Ordered TPN solution and rate
  1 liter/day = 43 cc/hour
  2 liter/day = 86 cc/hour
  3 liter/day = 125 cc/hour
- Ordered fat emulsion solution, if applicable
- Check solution label carefully against physician's orders for correct dextrose solution, electrolytes, and trace minerals Call pharmacy for any solutions that are not correct and obtain correct solution prior to hanging
- Check solution for precipitate; do not hang if present
- Begin TPN slowly usually at 25-50 cc/hour to assess for glucose tolerance problems

During TPN administration, monitor:

- Vital signs every 4-8 hours depending on agency policy
- Weight, at least every other day and assess for changes
- Clinitest and acetone may be ordered 3-4 times per day to assess for glucose in the urine. Report any reading of 2+ or greater on C&A or any serum blood glucose elevation so that insulin can be added or dextrose decreased
- Dressing should be checked each shift for intactness and changed as needed or daily as per agency policy
- Tubing should be changed daily or as per agency policy

- Strict aseptic technique should be used at all dressing and tubing changes to prevent sepsis especially in central lines
- Potassium levels for hyperkalemia, which is the most common electrolyte problem in TPN especially with elderly or renal patients. Report abnormal findings
- For metabolic alkalosis due to the conversion of synthetic amino acids to bicarbonate, report abnormal laboratory findings.
- Sodium level for hyponatremia, which may occur with TPN. Report any abnormal findings.
- TPN should not be discontinued suddenly, the patient should be weaned from the solution per physician's orders.

Chart:

- Date TPN begun
- Contents of TPN solution and rate of flow
- Current weight, admission weight, desired weight
- All C&As done on urine, with time
- Laboratory values:
  Glucose (detect glucose intolerance)
  BUN, creatinine (monitor rend function)
  Total protein, albumin, bilirubin
  Calcium, electrolytes (assess deficiencies or excesses)
  Cholesterol, triglyceride levels
- Times of rate change and new rate

Fluid Needs

Water is the most vital nutrient the body requires. The primary functions of water in our body include:

- Solvent to transport nutrients to cells
- Solvent to remove waste products from cells
- Solvent for electrolytes
- Lubricant for movement of joints
- Lubricant for food in mouth (saliva)
- Aid in chemical breakdown of foods (carbohydrates to monosaccharides, proteins to amino acids)
- Regulation of body temperature to prevent overheating (sweat)
- Promoting digestion and elimination
- Component of body secretions. The average healthy adult has between 55 percent and 65 percent of total body weight in water or about 42,000 ml or 175 cups. Water is located in several compartments of the body as shown in this table.

			% of body weight			
Area	Definition	Amount	Infant	Female	Male	Elderly
Intracellular (ICF)	Water inside cells	28,000 ml 117 cups	48%	35%	45%	25%
Extracellular (ECF)	Water outside of cells	14.000 ml 58 cups	29%	15%	16%	20%
Interstitial part of ECF	Water that transports nutrients/waste and bathes cells	10,500 ml 44 cups	25%	10%	11%	15%
Intravascular part of ECF (plasma)	The fluid in blood outside cells	3,5000 ml 14 cups	4%	5%	4%	5%
Total % of body weight in water			77%	50%	60%	45%

Now that we know roughly how much water is in the body we can look at how much of the total 42,000 ml is lost per day and how. On average, the normal healthy adult gains and loses about 2,400 ml per day.

Water losses in adult body

Amount/day	Route
800-1500 ml	Urine
250-350 ml	Stool
100-250 ml	Perspiration
250-350 ml	Skin
350 ml	Lungs
1900-2800 ml	Total lost

Water gains in adult body

Amount/day	Route
1000-1250 ml	Liquids
625-1250 ml	Solid food
200-400 ml	Metabolic oxidation
1825-2900 ml	Total gain

Calculation of Fluid Per Day Needs

To calculate the amount of fluid needed by your patient in a 24 hour period you can use either of these methods:

Allow 1 liter for every 1000 kilocalories in adult diet

or

Allow 100 ml/kg for first 10 kg of body weight

Plus 50 ml/kg for the next 10 kg of body weight

After the first 20 kg then use 20 ml/kg (persons less than 50)

After the first 20 kg then use 15 ml/kg (persons older than 50)

Example problem:

#l -- 15 kg child would require how much fluid/day?

100 ml x 10 kg = 1000 ml
50 ml x 5 kg = 250 ml
1250 ml/day

#2 -- 60 kg, 35-year old women would require how much fluid/day?

100 ml x 10 kg = 1000 ml
50 ml x 10 kg = 500 ml
20 ml x 40 kg = 800 ml
2300 ml/day

#3 -- 85 kg, 62-year old male would require how much fluid/day?

100 ml x 10 kg = 1000 ml
50 ml x 10 kg = 500 ml
15 ml x 65 kg = 1075 ml
2575 ml/day

Symptoms of fluid imbalance are important for health care workers or others to recognize because either hypovolemia (too little fluid in the body) or hypervolemia (too much fluid in the body) can be life-threatening. Symptoms of these two conditione are:

Hypovolemia	Hypervolemia
Dry skin	Edema
Dry mucous membranes	Puffy eyelids
Decreased urine output	Moist lung crackles
Weight loss	Weight gain
Fatigue	Dyspnea
Tachycardia	Bounding pulse
Tachypnea	Shortness of breath
Elevated RBCs	Decreased RBCs
Elevated hemoglobin	Decreased hemoglobin
Elevated hematocrit	Decreased hematocrit

Fluids are generally replaced during hospital stays using the following breakdown:

7-3 shift - 3/6 of the total fluid intake is given

3-11 shift - 2/6 of the total fluid intake is given

11-7 shift - 1/6 of the total fluid intake is given

If the last example above is used, we can calculate how much fluid this man would receive each shift. His total daily intake is 2575 ml per day.

7-3 shift, 3/6 = 1288 ml (approximate)
3-11 shift, 2/6 = 858 ml
11-7 shift, 1/6 = 429 ml
2575 ml

To force fluids on this patient, the amount of fluids would be doubled. His fluid intake would be 5150 m/day.

Intravenous (IV) Fluid Replacement

IVs are common among hospitalized patients to treat fluid imbalances. The common solutions are listed below.

Solution	KCal/L	Indications	Osmolarity	Na	Cl	K	Ca	HCO_3
D_5W	170	Prevent dehydration Promote diuresis of Na	Isotonic	--	--	--	--	--
$D_{10}W$	340		Hypertonic		--	--	--	--
$D_{20}W$	680	Hypoglycemia	Hypertonic	--	--	--	--	--
NS 0.9%	--	Diabetic acidosis	Isotonic	154	154	--	--	--
½ NS 0.45%	--		Hypototonic	77	77	--	--	--
$D_5$4¼NS	170	Fluid maintenance	Hypotonic	34	34	--	--	--
D_5½NS	170	Fluid loss, NA loss Promote diuresis	Hypotonic	77	77	--	--	--
D_5NS	170		Isotonic	154	154	--	--	--
Lactated Ringer's	Less than 10	Metabolic acidosis Dehydration Burns Infection	Isotonic	130	109	4	3	27
D_5 Lactated Ringer's	180	Diarrhea Loss of bile or pancreatic juice	Isotonic	130	109	4	3	27
Ringer's solution		Fluid maintenance	Isotonic	147	156	4	4.5	

Remember:

For patients on ventilators do not count losses from lungs/skin as this is compensated by humidified air.

For elevations in temperature each degree above 37°C will increase water loss by 2.5 ml/kg/day.

Patients in congestive heart failure or renal failure have decreased water intake due to fluid retention.

Every 240 ml is equal to one 8-ounce cup.

NS, NaCl, 0.9% are other names for mormal saline solution.

IV replacement of fluids does not meet the caloric needs.

D_5W contains 50 grams/liters of dextrose, D_{10} 100 grams/liter.

Ringer's solution and lactated Ringer's are not the same.

Calculating Drip Rates

In order to calculate the IV rate you must start with the physician's order.

Example order:

1000 cc D_5/0.45 NS to run at 120 cc/hr.

If you were given these orders, what would you do?

#1 -- Find out the drop factor of the tubing. Look on the tubing box for this information. Some of the more common drop factors are 10 gtts/min., 15 gtts/min., or a micro drop is 60 gtts/min.

#2 -- Once you know the drop factor you can calculate the rate by using this formula

$$\text{Drops/minute} = \frac{\text{amount to be infused}}{\text{minutes to infuse}} \text{ X drop factor}$$

Let's try this problem:

$$\text{Drops/minute} = \frac{120 \text{ X } 10}{60}$$

$$\text{Drops/minute} = \frac{\text{Amount to be infused (120 cc)}}{\text{Minutes to infuse (60)}} \text{ X drop factor (10 gtts)}$$

Drops/minute = 120 X 10 ÷ 60 =

Drops/minute = 20

Caloric Calculation

Examples of practices that interfere with nutrition include:

- Prolonged NPO status for tests, surgical procedures, or for patients with GI disorders (pancreatitis)
- Prolonged IV fluid supplement (low in calories and other needed nutritional needs
- Loss of appetite associated with treatments (radiation, chemotherapy)
- Increased nutritional need during physical stress (injury/illness)

Nursing intervention to prevent these problems:

- At admission, record the height and weight of every patient
- Periodically weigh every patient to assess for undesired weight loss
- Consult with the physician and dietician regarding any patient that has lost unplanned weight since admission to prevent further delay of nutritional support
- Record the percentage of food ingested after each meal
- When indicated, order meal trays missed due to diagnostic testing when patient returns
- Calculate the amount of calories in tube feedings and compare to calories needed per day for your patient
- Monitor and record all laboratory tests performed to assess nutritional status and inform physician of abnormal results

Calculation of Caloric Needs, Using BEE:

BEE stands for basal energy expenditure and is measured in calories per day. The following formula is used to calculate the BEE for adults.

Formula for Males:

66 + (13.7 x usual weight in kg) + (5 x height in cm) - (6.8 x age)

Formula for Females:

655 + (9.6 x usual weight in kg) + (1.7 x height in cm) - (4.7 x age)

Or you can estimate the BEE using this formula:

Male BEE = 1 Cal/kg/Hr.

Female BEE = 0.9 Cal/kg/Hr.

Remember when calculating:

- Weight in kg is calculated from pounds
  1 kg = 2.2 pounds

- Height in cm is calculated from inches
  2.54 cm = 1 inch
- If weight is remarkable for being over or under, calculations should be made on desired weight (formula given below)

Example problem:

Calculate BEE for a 28-year-old male who is 5'9" and weighs 170 pounds.

Convert 5' 9" into cm:

Ht (5 feet 9 inches) = (5 X 12) + 9 inches

Ht = 60 + 9 inches

Ht = 69 inches

Ht in cm = 69 X 2.54

Ht in cm = 175.26 cm, rounded to 175 cm

Convert 170 pounds into kg:

Wt in kg = $\frac{\text{Wt in pounds}}{2.2}$

Wt in kg = $\frac{170}{2.2}$

Wt in kg = 77.2 rounded to 77 kg

Now to calculate BEE, we can put the information into the formula:

28-year old male, 77 kg, 175 cm

BEE = 66 + (13.7 X 77) + (5 X 175) - (6.8 X 28)

BEE = 66 + (1054.9) + (875) - (190.4)

BEE = 1120.9 + 875 - 190.4

BEE = 1995.9 - 190.4

BEE = 1805.5 calories

In our example 1805.5 calories are required to sustain a 28-year-old man body at rest without physical stress.

To calculate the BEE of a person who is either overweight or underweight we may not want to use their usual weight which would give too many or too few calories for them to attain their desired weight. For these persons, you can obtain desired weight from the weight table or calculate the desired weight as follows.

	Males	Females
1st 5 ft. of height	106 lbs.	100 lbs.
For every inch over	6 lbs.	5 lbs.

This simple formula is for a medium-framed person.

10% of total body weight is added for large frames.

10% of total body weight is subtracted for small frames.

Using our example and stating he is a large-framed male, we can calculate his desired weight as follows:

Desired wt. (lbs.) = 106 + (9 X 6)

Desired wt. = 106 + 54

Desired wt. = 160 lbs. for medium frame

To adjust for large frame we add 10% of this total or 16 lbs.

Desired wt. = 160 + 16

Desired wt. = 176 lbs.

Our 28-year old male was actually 6 pounds underweight.

VITAMINS

A vitamin is an organic compound derived from the diet needed in very small quantities to promote metabolic processes responsible for growth and maintain life.

The following tables provide information on 13 major vitamins.

RDA is the Recommended Dietary Allowance
Antagonistic oppose the action of the vitamin
Synergist is a co-worker or helping factor for the vitamin

The first four vitamins are all fat soluble and are insoluble in water. These are A, D, E, and K.

Vitamin A

RDA	1,000 ug RE for adults
Principle sources	Egg yolk, liver, lamb chops, milk, butter Yellow-orange pigment foods: apricots, cantaloupe, peaches, carrots, sweet potatoes yellow squash Green vegetables: broccoli, spinach
Function in body	Night vision, mucous secretions, formation of cartilage, growth and repair of cells, needed for spermatogenesis
Storage in body	Liver
Overdose symptoms	Greater than 100,000 ug/day for 2-3 weeks Lethargy, abdominal pain, headache, sweating, joint pain, jaundice and loss of hair
Deficiency symptoms	Night blindness, spots, xerophthalmia, blindness, UTI, diarrhea
Persons at risk	Elderly, infants, diabetics, alcoholics, smokers, drug addicts, and persons with hyperthyroidism
Antagonists	Air pollution, strong light, mineral oil, or increased protein intake
Synergists	Vitamins D, E, and C
Comments	Carotene is the precursor to Vitamin A Fat soluble and insoluble in water and absorbed in small intestine

Vitamin D

RDA	10 ug/day for adults
Principle sources	Sunlight, liver oils, margarine, lard, egg yolks, Vitamins D milk, shrimp, salmon, tuna
Function in body	Growth and repair of bone Maintains balance of calcium and phosphorus
Storage in body	Precursors stored in skin
Overdose symptoms	Calcification of kidneys, blood vessels and skin 1,000 units/lb/day in adults, kidney stones
Deficiency symptoms	Rickets (bowed legs, growth retardation, rachitic rosary, pigeon breast) Osteomalacia, osteoporosis (fragile bones) Increased tooth decay, decreased muscle tone
Persons at risk	Pregnancy, areas with little natural sunlight, elderly, lead poisoning
Antagonist	Cortisone, anticonvulsant
Synergists	Vitamin A, B_1, B_3, C and calcium
Comments	Mild bone deformities due to diet may be reversed Serious deformities are permanent

Vitamin E/Tocopherol

RDA	8-10 mg a-TE/day
Principle sources	Margarine, oils, chocolate, peanuts, whole grains, wheat germ, yeast, asparagus, broccoli, cabbage
Function in body	Maintains cell membrane integrity Acts as an antioxidant
Storage in body	Muscle, fatty tissue and liver in small quantities

Overdose	Rare, may raise blood pressure
Deficiency symptoms	None found, except in low birth-weight infants who may suffer from RBC hemolysis
Antagonists	Rancid fats and oils, iron if taken at the same time, oral contraceptives, thyroid hormone, mineral oil
Synergists	Vitamin A, B complex, C, hormones, cortisone, testosterone, STH
Comments	Never take Vitamin E and iron together

Vitamin K

RDA	10 ug/day
Principle sources	Green leafy vegetables, cabbage, kale, spinach, alfalfa, beef, pork, liver, cauliflower, tomatoes, carrots
Function in body	Aids in synthesis of blood clotting factors
Storage in body	Liver in small amounts Vitamin K is produced by intestinal flora
Overdose symptoms	Not found, most likely due to limited storage
Deficiency symptoms	Rare in healthy persons; Hemorrhaging, hypoprothrombinemia or a tendency to bleed easily
Persons at risk	Newborn infants (until intestinal floras take over in first 72 hours) Chronic diarrhea, colitis or other disorders that interfere with intestinal absorption
Antagonists	Anticoagulants, penicillin, tetracycline, sulfonamide, aspirin, mineral oil
Synergists	Vitamin A, C, E
Comments	Vitamin K is often given to newborns

The water soluble vitamins include all of the B complex vitamins and Vitamin C. Some characteristics these vitamins share are:

Dissolve in water
Rapid absorption into body tissue
Rapid excretion from body

Need for regular dietary supplement due to rapid excretion
Poorly stored in the body
Excessive amounts are excreted in urine and sweat
Tend to work together (as co-enzymes) within the body
Easily destroyed by processing, storage, food preparation

Vitamin B_1/Thiamin/Thiamine

RDA	1.5 mg/day adult dose
Principle sources	Brewers yeast, wheat germ, whole grains, liver, kidney, pork, plums, prunes, raisins, sunflower seeds, nuts
Function in body	Maintain the health of nerves, heart, muscle and GI tissue, promote growth and cell repair, needed to release energy from carbohydrates
Deficiency symptoms	Fatigue, weight loss, stomach upset, weakness, memory loss, depression, irritability, Beriberi with above plus tissue swelling, mental and motor dysfunctions
Persons at risk	Cardiac, patients, alcoholics, elderly, allergies, post-surgical patients, temperature elevation
Antagonists	Emotional stress, physical stress, nitrates, baking soda, air pollution, alcohol, antibiotics
Synergists	Vitamins B_2, B_3, B_8, B_{12}, C, E, and pantothenic acid
Comments	Absorbed in the small intestine Minimize water in preparation of foods high in B_1

Vitamin B_2/Riboflavin

RDA	1.7 mg/day for adults
Principle sources	Liver, heart, kidney, milk, eggs, eggnog, broccoli, asparagus, brewer's yeast

Function in body	Needed for conversion of protein to energy Maintenance of mucous membranes Fetal growth and development
Deficiency symptoms	Fatigue, stomach upset, anxiety, personality disturbance, high blood pressure Angular fissure and cheilosis of lips Glossitis with smooth tongue Eyes sensitive to light, watery eyes
Persons at risk	Elderly
Antagonists	Antibiotics, oral contraceptives
Synergists	Vitamins A, B_3, E
Comments	The more protein in diet the greater need for B_2 Milk in clear glass bottles has less B_2, than milk in other packaging

Vitamin B_3/Niacin/Niacinamide/Nicotinic acid/Nicotinamide

RDA	19 mg NE/day for adults
Principle sources	Liver, chicken, turkey, tuna, halibut, swordfish, roasted peanuts, yeast
Function in body	Metabolism of fats, carbohydrates and proteins Body growth rate May reduce cholesterol and triglycerides
Deficiency symptoms	Pellagra (dermatitis, dementia, diarrhea) Lassitude, mild skin rash, irritability, headache, loss of memory, loss of appetite, darkening of skin
Persons at risk	Alcoholics, diabetics, cancer, colitis patients
Antagonists	Emotional stress, physical stress, antibiotics
Synergists	Vitamins A, B_1, B_2, B_6, B_{12}, C, D

Comments	Peak need is between ages 15-22 years (males), 7-14 years (females) About 50% of RDA comes from tryptophan being converted to B_3 by intestinal flora

Vitamin B_8/Pyridoxine/Pyridoxal/Pyridoxamine

RDA	2 mg/day for adults
Principal sources	Liver, meats, herring, mackerel, salmon, peanuts, walnuts, bananas, brewer's yeast, soybeans, lima beans
Functions in body	Needed for metabolism of proteins and fats Regulates synthesis of fats Aids in synthesis of protein Aids in formation of RBCs, bile salts
Deficiency symptoms	Skin sores, depression, irritability, weight loss, convulsions, confusion, anemia
Persons at risk	Hyperactivity, anemia, use of oral contraceptives, pregnancy
Antagonists	Oral contraceptives, isoniazid, cortisone, penicillamine
Synergists	Vitamins B_1, B_2, B_3, C, magnesium
Comments	Up to 30 mg/day may be required for women who use oral contraceptives

Vitamin B_{12}/Cobalamin/Cyanocobalamin

RDA	2.0 ug/day for adults
Principle sources	Liver, kidney, crab, salmon, sardines, herring, oysters, egg yolk, milk, cheese
Function in body	Needed by all cells Regulates the formation of RBCs Aids in synthesis of DNA, RNA Aids in synthesis of proteins and fats
Deficiency symptoms	Pernicious anemia (weakness, fatigue, glossitis, cracked lips) Memory loss, paranoia, moodiness

Persons at risk	Vegetarians, faulty absorption of cobalamin Lack of intrinsic factor
Antagonists	Aspirin, codeine, neomycin, oral contraceptives
Synergists	Vitamin B_1, Folic acid, biotin, pantothenic acid Vitamin A, C, E
Comments	Vitamin actually contains an atom of cobalt

B Complex/Folic Acid/Folacin/Folate

RDA	200 ug/day for adults
Principle sources	Kidney, liver, heart, spinach, asparagus, mushrooms, tuna, bran, yeast
Function in body	Needed by all cells Aids in synthesis of DNA, RNA, choline Regulates development of nerve cells (embryo/fetus) A component of genes and chromosomes Natural analgesic (painkiller)
Deficiency	Macrocytic anemia, diarrhea
Persons at risk	Alcoholics, sprue, pellagra, leukemia, pregnancy, intestinal disorders, oral contraceptives
Antagonists	Alcohol, emotional stress, physical stress, sulfonamides, methotrexate
Synergists	B-complex vitamins, vitamin C, hormones: testosterone, estradiol
Comments	Synthesized by intestinal bacteria in small amounts in normal healthy person

B Complex/Pantothenic Acid

RDA	10 mg/day for adults
Principle sources	Found in almost every food Brewer's yeast, meats, whole-grain cereals, bacteria of GI tract

Function in body	Aids in synthesis of cholesterol and steroid hormones Aids in energy release from carbohydrates, fats and proteins Needed for the synthesis of fatty acids
Deficiency symptoms	Very unusual to have deficiency Depression, fatigue, constipation, headaches
Persons at risk	Surgical patients or others with wounds High physical or emotions stressors
Antagonists	Methyl bromide (insecticide)
Synergists	B complex vitamins, Vitamins A, C, E, calcium
Comments	Known as the anti-stress vitamin

B Complex/Biotin

RDA	0.3 mg/day
Principle sources	Egg yolk, pork, liver, chicken, cereals, yeast, wheat, corn, milk, legumes, nuts, mushrooms, bacteria of BI tract
Function in body	Regulates metabolism of unsaturated fats Needed for normal function of sweat glands, nerve tissue, blood cells, male sex glands
Deficiency symptoms	Deficiency highly unlikely
Persons at risk	None found
Antagonists	Raw egg white (greater than 24/day), antibiotics, sulfonamide
Synergists	B complex vitamins, vitamin A, D, hormone STH
Comments	Intestinal flora synthesize biotin

Vitamin C/Ascorbic acid/Ascorbate

RDA	60 mg/day for adults

Principle sources	Citrus fruits, strawberries, cantaloupe, green peppers, broccoli, brussels sprouts, parsley
Function in body	Aids in fighting infection Aids in maintenance of blood vessel walls Promotes the absorption of iron Needed for formation of bone, teeth, and cartilage Aids in metabolism of some amino acids
Deficiency symptoms	Scurvy (tenderness in calves, muscular weakness, poor appetite, bleeding gums, loose teeth) Easy bruising, edema, poor wound healing
Persons at risk	Infants on unsupplemented formulas High levels of physical or emotional stress Wounds or surgical operations
Antagonists	Air pollution, smoking, alcohols, aspirin, diuretics, prednisone, indomethacin, steroids, antidepressants, anticoagulants
Synergists	B complex vitamins, vitamin A, K, E, hormones STH, testosterone
Comments	No evidence of cure for common cold

Nasogastric Tubes

Nasogastric tubes or NGs are indicated for:

Gastric decompression (suction)
Intestinal obstruction, post-op patients, ileus
Gastric gavage (feeding)
Inability or refusal to swallow or eat
Gastric lavage (washing out)
Drug overdose, gastrointestinal bleeding

Types of NG Tubes

Name	Primary purpose	Comments
Cantor tube	Intestinal decompression	Single-lumen with balloon tip filled with mercury so balloon will descend into small bowel Do not tape to nose or the tube will not descend
Keofeed	NG tube feedings	Weighted tube to travel into duodenum to prevent regurgitation
Levin	NG decompression	Single-lumen with perforated tips; 10-18 French
Miller-Abbott	Intestinal decompression	Double-lumen with balloon tip filled with mercury so balloon will descend into small bowel One lumen for aspiration, one for mercury Do not tape to nose
Salem-Sump	Gastric lavage, NG decompression	Double-lumen, small lumen is air intake vent Feedings for lavage
Sengstaken-Blakemore	Control bleeding Esophageal varices	Triple-lumen, one for aspiration, one for gastric balloon, one for esophagal balloon

Prior to Insertion of NG Tube:

- Check order for gastric intubation and purpose
- Select tube type if not ordered
- Select appropriate size (usually 16-18 F in adults)
- Assemble all equipment:

NG tube
Lubricant (water soluble)
50 cc irrigating syringe
Stethoscope
Suction equipment or feeding solution with primed tubing
Emesis basin, tissue, towel, adhesive tape, gloves

To Insert NG Tube:

- Position in high-Fowler's if possible
- Instruct patient in procedure
- Place towel on patient's chest, give basin to patient
- Measure tube from nose to earlobe to xiphoid process, mark with adhesive tape
- Lubricate tip (3-4 inches)
- Gently and steadily insert tube along nasal passageway floor
- If resistance is met, ask the patient to swallow small amount of cold water through a straw while you slowly continue
- When adhesive mark is at nose, stop and aspirate with syringe for gastric contents to assure you are in stomach. Another test for placement is to instill 15 cc of air while listening with stethoscope for rush of air over gastric area
- Tape to nose
- Begin suction, gavage or lavage as ordered
- Chart the procedure

Care of NG Tube After Insertion:

- Rotate insertion site from nostril to nostril to prevent tissue damage from pressure necrosis
- Provide meticulous mouth care to decrease dryness and irritation
- Petroleum jelly may be supplied at nares insertion site to decrease irritation
- Throat lozenges or spray may help with irritated sore throat
- Check for proper placement before each gavage feeding or restarting decompression
- Check for position at least once per shift on continuous feed or suction

There is a wide variety of feedings available (both NG and gastrointestinal feedings). A small sample is given below with the nutritional value of such formulas. The calories each supplies are especially important for patients on tube feedings.

Types of Formulas

The following figures are per 1 cc

Name	Calories	Carbohydrate	Protein	Fat	K+mg	Na=mg
Ensure	1.06	144.5 mg	7.1 mg	37.1 mg	1.27	.74
Ensure plus	1.50	199.0 mg	55.5 mg	53.3 mg	1.9	1.06
Osmolite	1.06	143.0 mg	37.0 mg	38.0 mg	0.875	0.541
Sustacal	1.00	148.0 mg	65.0	24.0 mg	2.2	1.0
Vital	1.00	185.0 mg	41.6 mg	10.2 mg	1.164	0.38

Complications of Tube Feedings:

As with any treatment tube feedings do have some complications. The two major complications are diarrhea and nausea. The causes are listed.

Complication	Cause
Diarrhea	Bacterial contamination Feeding too quickly Feeding too cold Allergic reaction Too much lactose Osmolarity too high
Nausea	Delayed gastric emptying

Section IV -- DIETS

Hospital diets are designed to serve the needs of patients who must alter their diet for physical or psychological reasons in order to restore or maintain their health. The regular or general diet is used for all patients who have no dietary restrictions. The following tables supply a brief description of other diets

Mechanical Soft Diet/Dental Soft Diet

Indications	Difficulty chewing (poor dentition, weakness) Difficulty swallowing (CVA, surgical procedure, radiation therapy) Step between liquid and general diet
Comments	Modification of regular diet Individualized, based on patient's needs No restriction on seasoning No restriction on food preparation
Restrictions	Based on individual assessment Difficult chewing raw vegetables and some fruits Foods with nuts, seeds Difficult swallowing Pulp fruits (prunes, plums, strawberries) Onions, custards, puddings, crackers Milk products (leads to increase in saliva)
Allowable foods	Chopped or pureed food Ground or pureed meats Pureed vegetables Liquids of all types Soups Foods with moist consistency Casseroles Soft foods Canned fruits Well-cooked vegetables Tender meats Baked chicken, turkey or roast beef
Other interventions	Offer food in bite-sized amounts Textured foods will stimulate swallowing:

	Toasted bread, crisp bacon, baked potatoes Position patient in high-Fowler's position

Clear Liquid Diet

Indications	Pre-Operative Post-Operative Step #1 from NPO to regular diet Required for some laboratory tests
Comments	Little nutritional value Provides fluids and relieves thirst
Restrictions	Solid foods Milk products Carbonated sodas (check agency policy) Fruit juices with pulp
Allowable liquids	Beverages Water, coffee, tea, lemonade, clear fruit juices (apple) Soups, clear broths, bullion, consomme Plain gelatins Hard, clear candies
Other interventions	Offer fluids frequently Pull privacy curtain if roommate has a more advanced diet during meals

Full Liquid Diet

Indications	Step #2 from NPO to regular diet Moderately-impaired GI function
Comments	More satisfied feeling after meals than with the clear liquid
Restrictions	Solid foods
Allowable liquids	All from Clear Liquid Diet plus: Beverages Strained fruit and vegetable juices Milk, milk-shakes, carbonated sodas Any of the usual beverages Soups Creamed soups, strained soups, gruels Dessert Ice cream, ices, custards, puddings

Low Residue

Indications	Indigestion, gastric reflux colitis, ileitis, diarrhea Radiation therapy
Comments	Eliminates high fiber foods Foods are generally soft Reduce GI discomfort
Restrictions	Milk and milk products All fried foods Highly-seasoned foods Nuts and seeds Coarse cereals or whole grain breads Raw vegetables, tough-skinned vegetables Corn, onions, dried beans, peas, potato skins Raw fruits (may have bananas) Jams with berries, seeds or skins
Allowable foods	Meats: Lean beef, chicken, lamb, turkey, veal, eggs Bread/Cereals: Refined flour, breads and rolls, rice, noodles, macaroni, spaghetti, and cooked cereals Vegetables: Cooked vegetables Fruits: Juices, stewed or canned fruits Fats: Butter, margarine, creams, oils, gravies Soups and desserts: Any without restricted foods
Other	Fats and seasonings may have to be eliminated if not tolerated by the patient

High Fiber

Indications	Constipation, diabetes, diverticular disease Irritable bowel syndrome Prevent intestinal disease
Comments	High fiber diets will stimulate peristalsis Decrease straining at stool especially when combined with increased fluid intake and activity
Restrictions	Refined foods and grains are limited

Allowable foods	All food groups allowed with increase in undigestible carbohydrates Foods high in fiber include: Wheat bran, wheat germ, whole wheat bread, nuts, cornmeal, raw fruits and vegetables
Other interventions	Encourage increase in activity Encourage 6-8 glasses of water daily Prune juice, while not high in fiber, contains a natural laxative agent

Reduced Calorie/Reduction Diet

Indications	Overweight or maintenance of desired weight
Comments	Caloric level will vary based on amount of need from 800-1600 calories/day Success rate is higher when exercise is begun or increased
Restricted foods	High fat foods: Margarine, creams, gravies, oils High calorie desserts and sweets: Cakes, pies, ice cream, candy Fried foods: French fries, deep fried meats
Allowable foods	All of the basic food groups should be included with smaller servings
Comments	Encourage reduced portion sizes Encourage pre-planning with patient of daily menu with inclusion of all four food groups within the allowable caloric intake Artificial sweeteners may help reduce cravings for sugar and can be used in cooking/baking Skim milk and low-fat cheeses should be encouraged

Low Fat

Indications	Gallbladder, liver or pancreatic disease
Comments	Usually ordered as 30-50 gm of fat/day

Restrictions	Fried foods: Fried meats, potatoes, vegetables Fatty foods: Oils, margarine, shortening, lards High fat meats, sauces and gravies High fat content dairy products: Ice cream, whole milk, cheddar cheese
Allowable foods	Meats: 5 ounces daily, prepared without fat or gravy (baked, broiled, steamed, roasted) All fat must be removed Fats (limited strictly): Generally only 3 servings/day Serving 1 pat of margarine Fats eliminated from cooking Fruits, vegetables: No limitation as long as served without breaking restrictions Dairy products: Skim milk, low fat yogurt and cheeses
Other interventions	Patient may complain of dry, tasteless food Seasonings may help improve taste of foods

Section V -- DRUGS

The tables below supply only general information. A drug handbook should be consulted prior to administering any unfamiliar drug.

ANTACIDS

Action: Neutralize acid in the stomach.

Indications: For use in conditions with excess production of hydrochloric acid such a gastric or peptic ulcers, hiatal hernia, esophageal reflex and gastritis.

General comments: Administer between meals unless otherwise noted. Tablets should be chewed thoroughly and liquids shaken well. Liquid preparations are reported to be more effective than the tables. Encourage increased fluid intake when taking antacids. Assess for patterns of stools for constipation or diarrhea as both are common side effects. Do not administer antacids within 1-2 hours of other drugs to prevent interactions.

Examples of drugs in this classification:

Aluminum Antacids:

Generic	Trade	Comments
Aluminum carbonate	Basaljel	May be given PO Major side effect is constipation Administer one hour after meals/bed-time
Aluminum hydroxide	Alternagel Alu-cap Alu-Tab Amphojel Dialume	May be given PO, intragastric drip Impaction or intestinal obstruction may occur with use, monitor stools
Aluminum phosphate	Phosphajel	Monitor serum phosphate levels and report elevation

Calcium Carbonate:

Generic	Trade	Comments
Calcium carbonate	Alka-2 Tums	May be given PO: tablets or powder

		Acid rebound with gastric hypersecretion may occur with chronic use Discourage long term dependence Monitor stools for constipation Assess for hypercalcemia with: nausea, vomiting, thirst, dry mouth, constipation, fatigue, muscle weakness, joint pain

Magnesium Antacids:

Generic	Trade	Comments
Magnesium carbonate		May be given PO Laxative effect Monitor for diarrhea
Magnesium hydroxide	Milk of Magnesia	Given before meals/bedtime Laxative effect
Magnesium oxide	Mag-Ox Uro-Mag Maox	May be given PO Laxative effect

Sodium Bicarbonate:

Generic	Trade	Comments
Sodium bicarbonate	Soda Mint Bell/ans Baking soda	Often abused Powder should not be taken until bubbling stops
	Alka-Seltzer	Short acting antacid Composed of bicarbonate and citric acid

Combinations of Aluminum & Magnesium:

	Maalox 1 and 2 Aludrox Mylanta Gelusil Di-Gel	Side effects from either category many occur

ANTICHOLINERGICS

Action: To reduce intestinal mobility and reduce acid production.

Indications: For use in condition with excess intestinal mobility or hydrochloric acid production to include peptic ulcers, pylorospasm, irritable colon.

General comments: Common side effects include dry mouth, increased heart rate, constipation, blurred vision.

Examples of drugs in this classification:

Generic	Trade	Comments
Atropine sulfate		May be given PO, SC, IM, IV Usually given 30 min. before meals Monitor vital signs Monitor I&O for urinary retention Monitor stools for frequency Provide mouth care for dryness
Belladonna		May be given PO Do not administer with antacid Provide mouth care for dryness

ANTISPASMODICS

Action: Relieve muscle spasms in the GI tract, thereby relieving pain

Indications: Infant colic, gastritis, irritable colon where muscle spasms of the GI tract are a problem.

General comments: Usually given before meals or at bedtime, common side effects similar to those of anticholinergic, administer cautiously in infants

Examples of drugs in this classification:

Generic	Trade	Comments
Dicyclomine hydrochloride	Bentyl Antispas	May be given PO or IM (adults) Tablet may be crushed and food added May lead to reduced abiLity to perspire, therefore, excess heat should be avoided

ANTIDIARRHEAL

Action: To decrease the fluid volume in and the amount of loose stools.

Indications: Frequent loose watery stools, diarrhea.

General comments: Encourage fluid intake to prevent dehydration, during episodes of diarrhea, clear liquids should be offered and solid foods avoided, electrolytes should be monitored for imbalances and replaced as needed.

Examples of drugs in this classification:

Generic	Trade	Comments
Bismuth subsalicylate	Pepto-Bismol	May be taken PO Contains large amount of aspirin Tablets should be chewed or allowed to dissolve before swallowing Available over-the-counter Stools may appear black in color
Kaolin/pectin	Kaopectate K-Pek	May be taken PO Available over-the-counter
Lactobacillus	Bacid Lactinex	Used mainly for diarrhea from antibiotics Available over-the-counter
Loperamide hydrochloride	Imodium	Usually given after each stool up to 16 mg daily May be given PO May cause drowsiness/dizziness Used often for ulcerative colitis Activated charcoal for overdose
Diphenoxylate hydrochloride	Lomotil Elmotil	May be given PO Schedule V drug Tablets may be crushed

ANTIFLATULENTS

Action: To aid in the expulsion of gas.

Indications: For excess, uncontrollable intestinal gas

Examples of drugs in this classification:

Generic	Trade	Comments
Simethicone	Mylicon Silain	Chew tablets well, shake liquids well Use after meals and at bedtime May be given with antacids

DIGESTIVE ENZYMES

Purpose: Replace enzymes needed for digestion.

Indications: A lack of digestive enzymes, found in cystic fibrosis, obstruction of the bile ducts or pancreatic ducts

General comments: These drugs aid in the breakdown of fats, proteins, starch/antacids may be given before or with these drugs.

Examples of drugs in this classification:

Generic	Trade	Comments
Pancreatin	Pancreatin Viokase	May be given PO Enteric coated tablets should not be crushed
Pancrelipase	Cotazym Ilozyme Ku-Zyme HP Pancrease	Does related to dietary fat intake

EMETICS

Action: To induce vomiting

Indications: Ingestion of toxic substance or drug overdose

General comments: Prior to administration in suspected or known poisoning, poison control center should be notified of ingested substance as vomiting may be contraindicated

Example of drugs in this classification:

Generic	Trade	Comments
Syrup of Ipecac		May be taken PO Available over-the-counter Should be taken with water Do not administer with activated charcoal

HISTAMINE BLOCKERS

Action: To reduce gastric acid

Indications: Conditions where gastric acid needs to be reduced such as in gastric ulcers, duodenal ulcers, stress ulcers

General comments: Antacids are often used concurrently with histamine blockers to treat ulcers

Examples of drugs in this classification:

Generic	Trade	Comments
Cimetidine	Tagamet	May be taken PO, IV push, IV, IM Administer with or just after meals Administer antacids at least one hour before or after this drug Many drug interactions

Ranitidine	Zantac	May be taken PO Long term use may lead to B_{12} deficiency

LAXATIVES

Action: To aid in the passage of stool

Indications: Constipation, hard, infrequent stools

General comments: Laxative abuse is common, laxative in general should be taken with fluids. Encourage diet high in fiber, increased fluids, exercise, to promote normal bowel function without laxative. Laxative abuse can lead to dependence as the GI tract may lose ability to expel stool without medication.

Examples of drugs in this classification:

Bulk forming agents generally work in 12-24 hrs. but may take up to 72 hrs.

Generic	Trade	Comments
Methylcellulose	Maltsuprex Cologel Isopto-Plain	May be taken PO Do not chew or crush tablets Administer with at least 8 oz. of water
Psyllium	Metamucil Mucillium Mucilose Reguloid V-Lax	May be taken PO May be mixed with water, fruit juice, milk or any liquid Follow with 8 oz. of water for best results Some products are high in Na+

Fecal softener

Generic	Trade	Comments
Docusate calcium Docusate sodium	Pro-Cal Colace Dio-Sul Molatoc Regutol Stulex	May increase absorption of other drugs, leading to toxicity at lower doses Sodium product should be avoided in patients with kidney/heart problems Stool softened in 1-3 days

Hyperosmolar laxatives:

Generic	Trade	Comments
Lactulose	Chronulac	Works in 24-48 hrs May be given PO Avoid giving with meals due to taste Given to detoxify ammonia in hepatic disorders Avoid exposure to light
Magnesium Hydroxide	Milk of Magnesia Magnesium Citrate Citroma	Works in 2-6 hrs Avoid use in renal patients Shake well Give before meals or at bedtime

Lubricants:

Generic	Trade	Comments
Mineral oil	Agoral plain	Do not take with food or medication Take with fruit juice May be given PO, rectal route Usually administered in evening

Stimulant Irritant:

Generic	Trade	Comments
Bisacodyl	Dulcolax Nuvac Theralax	May be given PO Do not crush or chew tablets Do not take with milk: or antacids Works in 1 hr
Cascara sagrada		May change color of urine Usually taken in evening Works in 6-12 hrs
Castor oil	Alphamul Emulsoil Neoloid Purge	May be give with juice for taste Should be taken on an empty stomach Works in 3 hrs
Danthron	Dorbane Modane	May change color of urine Usually taken in the evening
Phenolphthalein:	Correctol Ex Lax Feen A mint Prulet	Effect may last 3-4 days Usually taken in the evening Works in 6-12 hrs May change color of urine Discontinue if rash develops

Section VI -- GLOSSARY

Accessory organ An organ that assists other organs to perform their functions

Amino acids The basic building blocks of protein

Antagonists Counteracts the action of something else

Ascites The accumulation of fluid in the peritoneal cavity

BEE Basal energy expenditure is the amount of energy required to maintain life at rest

Bitos spots Shiny gray spots on the conjunctiva, due to a vitamin A deficiency

Bolus A portion of chewed food ready to be swallowed

Borborygmi The loud rumbling, gurgling sound heard when gas is moved down the intestinal tract

Bruits Sound heard during auscultation, arises from an arterial or venous source; always abnormal

BUN Blood urea nitrogen, a blood chemistry test to detect urinary function

C&As Urine test to detect the presence of glucose and ketones in the urine; stands for Clinitest and Acetone

cal Calorie

Cheilosis Reddened lips with fissures at the angles due to deficiency of vitamin B complex

Chyme A mass of partially digested food with digestive enzymes, found in the stomach and small intestine during digestion

cm	Centimeter, a unit of measure, 1 in = 2.54 cm
Constipation	Dry, hard, infrequent stools that are difficult to expel from the body
$D_5W_{OE}D_{10}W$	Dextrose solution in water, available in a wide range of concentrations
Deglutition	The act of swallowing
Diabetes mellitus	A condition with either inadequate or lack of insulin production. Symptoms include polydipsia (increased thirst), polyphagia (increased hunger), and polyuria (increased urination)
Diarrhea	Loose, watery, and frequent stools
Digestion	The chemical and mechanical breakdown of food so that it can be absorbed
Disaccharides	A carbohydrate made up of two monosaccharide (simple sugars)
DNA	Deoxyribonucleic acid; genetic material
Edema	Swelling in the body tissues
Elimination	Removal of wastes from the body
Emesis	Vomit
Feces	Product of bowel elimination
Force fluids	To increase fluid intake, usually to twice the normal fluid requirement
Gastrointestinal	Pertaining to the stomach and intestinal tract
Gastric decompression	To remove pressure from the GI tract via suction, also removes gastric contents that are irritating
Gastric gavage	Provide feeding through nasogastric tube when feeding by mouth is not possible or refused

Gastric lavage	To wash out the stomach
GI	Gastrointestinal
Glossitis	Inflammation of the tongue that is characterized by redness, pain and swelling
GT tube	Gastrostomy tube, a tube surgically placed through the abdominal wall, used to provide feedings when a patient cannot or will not take food by mouth
Hemorrhoid	An external or internal dilated vein in the anal area
Hernia	The abnormal protrusion of a partial or whole organ through the wall of a body cavity
Hepatitis	Inflammation of the liver, may be accompanied by jaundice
ht	Abbreviation for height
Hyperkalemia	More potassium than normal in the bloodstream
Hypervolemia	Increased fluid in the bloodstream
Hyponatremia	Decreased sodium in the bloodstream
Hypovolemia	Decreased fluid in the bloodstream
Indigestible	Not digestible
Ingestion	The process of taking food into the body via the mouth
Jaundice	A sign of excess bilirubin in the blood, characterized by yellowness in the sclera of the eyes, skin and mucous membranes
K+	Symbol for potassium
kg	Kilogram, a unit of measure 1 kg = 2.2 pounds

Lipids	Fat or fatty substance that is insoluble in water
Mastication	Chewing
Monpaccharides	A simple sugar such as fructose, galactose or glucose
Na+	Symbol for sodium
NaCl	Symbol for sodium chloride
Nasogastric tube	A soft flexible tube passed from the nostril to the stomach or intestine
Necrosis	Death of an area of tissue that is generally caused by an insufficient blood supply
NG	Nasogastric, as in nasogastric tube
NPO	Nothing per os, or, nothing by mouth
NS	Normal saline or 0.9% sodium chloride
Nutrient	Substance necessary for life such as water, minerals, vitamins, carbohydrates, fats, proteins and electrolytes
Osmolarity	The ion concentration of a solution, isotonic solution being osmotically the same as tissue fluids
Osteomalacia	Softening of the bone caused by vitamin D deficiency in adults, symptoms may include pain in bones, anemia and weakness
Osteoporosis	Condition of Ca++ loss from the bone see in elderly population; results in porous or weak bone structure
Pellagra	A disorder due to the deficiency of niacin, with glossitis, stomatitis, diarrhea, mental confusion, and skin that is dry with dermatitis

Pepsin	An enzyme found in gastric juice that breaks down proteins into peptone and proteose
Peristalsis	Wavelike contractions of the gastrointestinal tract to move the products of digestion
Pureed food	Ground or blenderized food that is semisolid without lumps
RNA	Ribonucleic acid
Rx	Symbol for prescription
Salivary gland	The glands which produce saliva to lubricate food and start digestion
Scurvy	A disorder characterized by fatigue, anemia, weakness, bleeding gums and hemorrhage
Skin turgor	Normal tension found in well-hydrated healthy skin. Turgor is measured by pinching up a small area of skin. With good skin turgor, the skin returns immediately to position
Sprue	A disorder characterized by weakness, weight loss, and impaired digestion
Striae	A stretching of the skin's outermost layer, often seen in pregnancy, and is known as "stretch marks"
Supine	To lie on the back
Synergists	A helper, as in vitamins, one that aids or potentiates functioning
Synthesize	To form a complex structure from smaller parts
Total parenteral nutrition	TPN, an intravenous solution that provides some essential nutrients and calories
Tube feedings	Feedings provided via the nasogastric or gastrostomy tube when normal ingestion of food is insufficient

Ulcers	A sore found in the mucous membranes (GI tract) or skin with inflammation and a gradual disintegration of surrounding tissue
Vesicles	A blister-like sac that contains fluid
Vitamin	An organic compound derived from the diet needed in very small quantities to promote growth and maintain life
wt	Abbreviation for weight
Xiphoid process	The lower tip of the sternum
Xerophthalmia	Dryness of the conjunctiva due to a deficiency of vitamin A

Reproductive System

REPRODUCTIVE SYSTEM

Table of contents

Section I -- OVERVIEW

Primary functions

The primary function of this system is the reproduction of the human species, to include the duplication of genetic information and the transfer of such information from one generation to another.

Components and function

The human reproductive system can be divided into two parts, the male and female reproductive systems. Primary components and functions of the male reproductive system are:

- To produce mature sperm cells capable of fertilization
- To store and transport sperm cells to the female reproductive system where the sperm may fertilize the female eggs
- The male reproductive system is composed of the testes, vas deferens, seminal vesicles, ejaculatory duct, the postrate and the penis. Figure 6A shows where the components are located

The testes

The testes are suspended by the spermatic cords and contained in the scrotal sac outside of the body. The location of the testes is related to spermatogenesis, or the production of sperm, which requires a temperature a few degrees lower than normal body temperature.

The left testicle is suspended slightly lower than the right one. The appearance of the scrotum changes under different conditions.

In older males or in warm temperatures the scrotum is longer. In young males or in cold temperatures it is short and close to the testes.

Vas deferens

The vas deferens (ductus deferens or seminal duct) is the excretory duct where semen leaves the testes and are transported to the prostatic urethra. It is palpable as a cord-like structure.

Seminal vesicles

Two small sac-like structures located behind the bladder and connected to the vas deferens. The seminal vesicles secrete a thick fluid which forms part of the semen. They do not store or secrete sperm.

Ejaculatory duct

The ejaculatory duct is about 2 cm long and begins at the end of the vas deferens. It begins where the seminal vesicles empty their contents into the vas deferens.

Penis

The penis is a cylindrical organ when flaccid, when erect, as during sexual arousal, it becomes more triangular in shape. The penis has a prepuce or foreskin that covers the glans penis. The penis contains the urethra which transports urine and semen.

Spermatogenesis and transport

Spermatogenesis takes place in the testes. Each sperm cell begins as a primary spermatocyte with 46 chromosomes. The primary spermatocyte divides into two secondary spermatocytes each with 23 chromosomes. From here the secondary spermatocytes divide again into four spermatids.

The spermatids have 23 chromosomes each and will develop into mature sperm with a head, middle piece and tail. The head contains the nucleus, genetic material and acrosome. The middle provides the energy for movements, while the tail is a flagellum that provides propulsion. During ejaculation the sperm are propelled via the vas deferens and mixed with fluid from the seminal vesicles and prostate to form semen. About 300 million sperm are produced daily.

Female reproductive system components and functions:

Primary functions of the female reproductive system are:

- To produce and store ova (eggs).
- To provide a medium for the fertilization of ova.
- To provide nutrients for and removal of waste products from the developing fetus during the 40-week gestation period.
- To transport the developed fetus from the uterus to the external environment via the vagina.
- The female reproductive system is composed of the ovaries, fallopian tubes, uterus, vagina, labia, and clitoris.

Figures 6B and 6C show where the components are located.

Ovaries

The ovaries are similar in structure to the testes. The ovaries are located internally one on either side and slightly above the uterus.

Unlike the male testes the ovaries contain a fixed number of primary oocytes or undeveloped eggs (from 100,000 to 400,000) present since birth. At puberty these ova will develop past the primary stage during oogenesis.

Fallopian tubes (oviducts)

The fallopian tubes will transport the ova from the ovaries to the uterus. The end of the tube near each ovary is funnel-shaped with

The Male Reproductive System

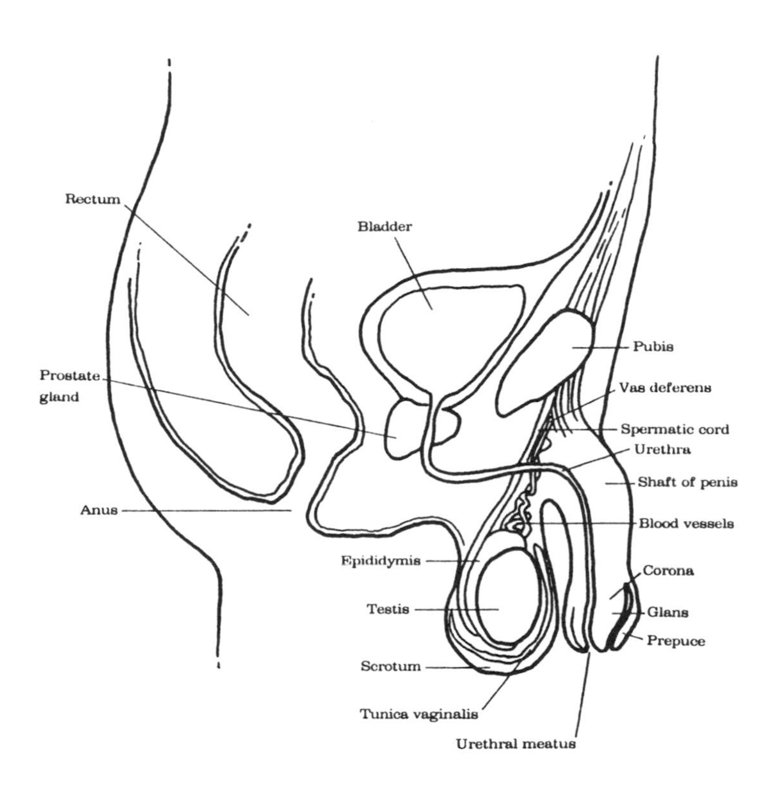

Figure 6A

The Female Reproductive System

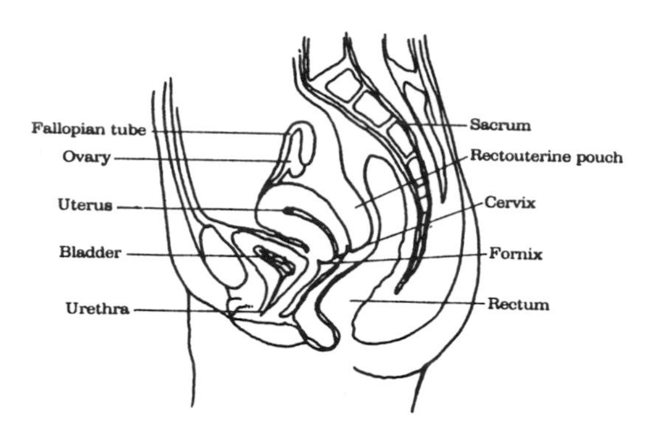

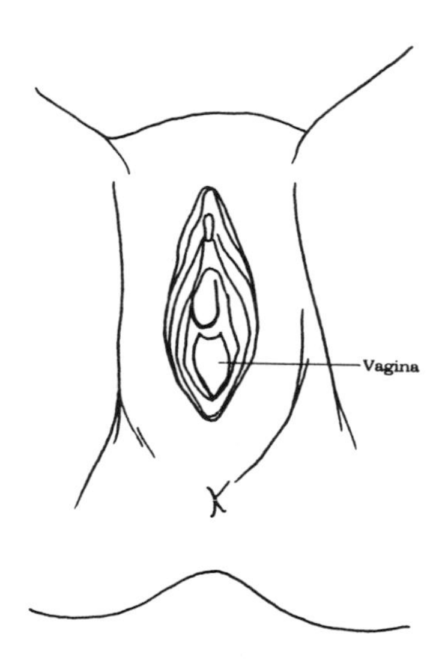

Figure 6B

fringed projections. The tube is lined with cilia which will propel the ova toward the uterus.

Uterus

The womb or uterus is a hollow organ between the bladder and rectum. At the top of the uterus the fallopian tubes enter on either side. At the bottom of the uterus is the vagina. The walls of the uterus are muscular and the uterus has the ability to change size for the developing fetus during pregnancy. The innermost layer of the uterus is shed during menstruation.

Cervix

The cervix is the end of the uterus that projects into the vagina. The opening, or os, of the cervix forms a path between the uterus and the vagina. During pregnancy the normally-firm structure becomes soft and pliable.

Vagina

The vagina is a muscular tube lined with mucus secreting cells. It is located behind the bladder and urethra. The vagina is approximately 7-10 cm in length and serves as a route for menstrual fluids as well as childbirth.

Labia

There are two sets of labia; the outer labia majora and the smaller, inner, labia minora. Between the labia minora are the openings for the urethra and vagina.

Clitoris

The clitoris is formed where the two labia minora join at the top. It is similar to the penis and during arousal will become engorged.

Oogenesis and transport

The female counterpart to spermatogenesis is oogenesis. The production of ova occurs in the ovary, the female counterpart to the testes. The primary oocytes were formed and stored before birth. At puberty the rising level of FSH (follicle stimulating hormone) stimulates the primary oocyte to divide into a secondary oocyte and a smaller polar body which will usually disintegrate. Both new cells have 23 chromosomes. During ovulation the secondary oocyte is released into the fallopian tube for possible fertilization. The oocyte undergoes division again only if fertilization occurs.

Fertilization and development of the product of conception

With the act of sexual intercourse semen containing sperm is expelled from the penis into the vagina. The sperm then make their way up the female reproductive tract. Fertilization usually occurs in the upper third of the fallopian tube (oviduct) during the oocyte's transport to the uterus. One mature sperm fertilizes one oocyte which then becomes the ovum. After fertilization, the ovum cannot be penetrated by other sperm. The fertilized ovum contains 23 chromosomes from the sperm and 23 from the ovum for a total of 46.

Rapid cell division follows, and the blastocyst moves down the fallopian tube into the uterus where it will attach (implant) to the inner wall 7-8 days after fertilization. The embryonic period will last two months and during this time the structures required to support fetal life and all primary adult organs will develop. At the end of the second month the embryo will be called a fetus.

During the last 7 months of uterine growth the fetus will continue to grow, develop and mature.

The male sperm determines the sex of the fetus. All oocytes contain an X chromosomes. If the sperm which fertilizes the ovum contain an X, the offspring will be female (XX); if the sperm bears the Y chromosome; the offspring will be male (XY).

Menstruation

When fertilization does not occur, menstruation results. At the onset of puberty (about 12-13 years) regular monthly menstruation begins and continues until menopause. Menstruation is the disposal by the body of its preparation for pregnancy. The menstrual cycle is composed of four phases:

- Menstrual phase lasting about 5 days
  Lining of uterus (endometrium) is shed; lining was prepared for implantation and to support embryo.
  Consists of a discharge of mucus, epithelial cells, blood and tissue.
  Total amount of fluid lost is about 25-100ml per month.
- Proliferative phase lasting about 10-13 days
  At start of phase inner lining of uterus (endometrium) is very thin.
  Ovarian follicles are stimulated by FSH and LH to produce more estrogen.
  The increased estrogen thickens the endometrium to prepare for the possible implantation of a fertilized ovum.
  Concurrently with this, an oocyte is being prepared for release from the ovary.
  Phase ends when the mature ovum is released at ovulation.
- Luteal phase lasting about 10-14 days
- Begins with the release of ovum at ovulation.

The corpus luteum secretes progesterone as well as estrogens which ovum increases the vascularization and thickening the endometrium. Progesterone secretion increases and is dominant. Once ovulation occurs the ovum has about 24 hours to be fertilized by the male sperm. If no fertilization occurs, implantation will not occur. Without implantation the corpus luteum will not be maintained.

- Premenstrual phase lasting about 1-2 days

  Without fertilization the corpus luteum degenerates and the endometrium is sloughed.

Section II -- ASSESSMENT

Assessment

It is often very difficult for someone to discuss problems of the reproductive system, and a supportive non-judgmental manner is important. Some individuals find it more comfortable to discuss sexual concerns with a health care worker of the same sex.

Assessment of this system will be divided into assessment of the male and female reproductive systems.

Male Health History

Chief complaint

Examples of common chief complaints:

Discharge from the penis
Burning on urination
Itching or rash in the genital area
A sore in the genital area that will not heal
Inability to retract the foreskin
A change in sexual desire or performance

Personal and family history

A positive history of past problems in the patient or cancer in a family member may help the physician in gathering information for a diagnosis.

Testicular cancer
Makes up 2 percent of cancers in men. Ranks second as the cause of death from all cancers in men between 20-34 years of age. Presenting symptoms include mass or edema in testicle, testicular pain, and sometimes enlargement of breast tissue

Prostrate problems
Benign prostatic hyperplasia (BPH) is commonly seen after age 50 and leads to urinary flow obstruction. Prostate cancer is another problem

Fertility problems
Sexually active without birth control and no conception to date. Infertility testing-low sperm count

Sexually transmitted diseases (STD)
Assess for type of STD, treatment, and compliance with treatment. Were both, or all, parties treated?

Current problems
Penile discharge, sores on genitals, lumps, scrotal pain, difficult or painful urination are all abnormal. Problems with sexual inter-

course can include impotence, premature or painful ejaculation or failure to ejaculate and lack of sexual drive

Reproductive System Testing

Sperm studies

Includes gross and microscopic examination, sperm count, and sperm morphology

Blood testing

Serum liver profiles, human chorionic gonadotropin (HCG) titer, alpha-fetoprotein (AFP) levels may be done in evaluation of testicular cancer. Blood testing may also be done to confirm STD such as syphilis

Current information

List all physicians who see the patient and all medications taken

Self-testicular examination

If self testicular exams are not performed, patient should be instructed in exam procedure and the importance of early detection

Sexually active with multiple partners

Inform patient of risk of HIV and STD infection Encourage use of condoms and one sexual partner to decrease risk

Female Health History

Chief complaint

Examples of common chief complaints:

Painful menstrual cramps
Premenstrual syndrome with depression, weight gain, anxiety
A cessation of menstruation (pregnancy or menopause)
A yellow or grayish vaginal discharge with burning
Itching or rash in genital area
A sore in the genital area that will not heal
A lump in the breast tissue
A small, painless lump or cyst on the side of the vagina
A change in sexual desire or performance
Personal and family history

A positive family history of breast or cervical cancer places the woman at greater risk

Cancer

The risk woman developing breast cancer are approximately one in nine, according to the American Cancer Society

Reproductive history

Total pregnancies are recorded as gravida and includes any abortions, miscarriages and children born

Problems may include bleeding, excessive emesis, preterm births, abruptio placenta, placenta previa, gestational diabetes, infant distress or delivery by caesarean section

Menstrual cycle

MP stands for menstrual period

LMP stands for last menstrual period

Last menstrual cycle for possible pregnancy or menopause. Difficulty in menstruation may be related to hormone levels, stress, fibroids, endometriosis, pelvic infections or IUD's

Vaginal discharge/burning/lumps/pain

Causes of abnormal vaginal discharge include bacterial vaginosis, candidiasis, chylamydia, pelvic inflammatory disease (PID) or trichomoniasis. A yellow or gray discharge is abnormal.

Pain, itching, or burning may be related to Bartholin cyst, candidiasis, chlamydia, genital warts, or trichomoniasis

Lumps in genital area may be due to Bartholin's cyst or genital warts

Reproductive system testing

Regular Pap smears are recommended for the early detection of cervical cancer

Mammograms should be done after age 35-40 for baseline

Current information

List all physicians seen and medication taken

Monthly self-breast examination

Instruct in importance of monthly breast self examination

Offer demonstration and assess patient's technique

Sexually-active with multiple partners

Inform patient of risk of HIV

STDs or PID can lead to infertility or sterility

Encourage monogamy and the use of condoms to decrease risk

Method of contraception

Should be assessed and if needed explanations of effectiveness of different options explored

MALE REPRODUCTIVE SYSTEM ASSESSMENT FORM

Chief complaint

Patient's statement________________Onset__________Symptoms________________

Frequency__________Duration__________Other areas affected__________________

Have you had this before?_________When?____________________________________

What have you done for this problem?______________________________________

What do you think caused this to happen?___________________________________

What changes have you had to make because of this problem?

____

What does this illness mean to you?____

Personal and Family History

	Patient Date	Family member		Patient Date	Family Member
Testicular cancer?	_____	_____	Fertility problems?	_____	_____
Prostate problem?	_____	_____	Sexually transmitted disease (STD)? (Name)____________		

Current problems:

Any unusual penile discharge?__________________Lumps in testicles?______________

Sores on penis or testicles?_________________Pain in testicles?___________________

Burning with urination?______________________Difficult urination?_______________

Problems related to sexual intercourse?____________________________________

Reproductive System Testing

Sperm studies______________________Blood tests_____________________________

Who is your physician?_______________________________Phone___________________

Are you sexually active?__________________How many partners?__________________

Do you use birth control or protection?_______What type?_________________________

Do you do self testicular-exams?_________________When_____________________________

Current medications:

Name_________________________Dose_________Frequency_________Route________

Name_________________________Dose_________Frequency_________Route________

Is there anything else you want me to know?_____________________________________

FEMALE REPRODUCTIVE SYSTEM ASSESSMENT FORM

Chief complaint

Patient's statement______________________ Onset________ Symptoms______________

Frequency__________ Duration__________ Other areas affected______________________

Have you had this before?__________ Date of last episode______________________

What was done for this problem?______________________________________

What do you think caused this to happen?______________________________

What changes have you had to make in your life because of this problem?

____

Personal History

Cancer of: Breast______________________ Total pregnancies ______________________

Uterus or cervix________________ Living______________ Deceased________________

Abortions/Miscarriages____

Problem in pregnancies or delivery?____________________________________

Age at MP onset?______________ LMP?____________ Problems with periods__________

Vaginal burning__________ Discharge__________ Unusual lumps________ Pain_______

Reproductive System Testing

Last pap smear?______________ Cervical biopsy?____________ Colposcopy____________

Mammogram?______________ Breast biopsy?____________ Other?________________

Who is your physician?________________________________ Phone______________

Are you sexually active?____________ How many partners?______________________

Do you use birth control or protection?____________ What type?__________________

Do you do self breast exams?______________ How often?______________________

Current medications

Name________________________________ Dose________ Frequency______ Route_____

Name________________________________ Dose________ Frequency______ Route_____

Is there anything else you want me to know?______________________________

Physical Assessment

Physical assessment of this system is done in this order:

1. Inspection
2. Palpation

Assessment of this system may be embarrassing for some patients to help promote relaxation and reduce anxiety:

- Explain what you will be doing before the examination
- Have patient empty bladder
- Wear gloves and use gentle touch during examination
- Patient will need to be undressed and gowned. Male patients are examined standing, female patients on exam table
- Expose only the area needed for the examination; use drapes

The forms on pages 269 and 270 can be used for physical assessment

Male Physical Examination

Inspection and palpation:

Breast
Abnormal findings: asymmetry, dimpling and discharge. Men can acquire breast cancer, palpate for masses

Sexual maturity
Inspection only, no palpation. Use sexual maturity scale rating system such as the one developed by Tanner

Pubis
Pediculosis pubis is crab lice infestation
A rash in the pubic area may be present with STD or fungal infection
Size, location of any hernias/masses should be noted. Ask the patient if he can reduce the hernia by position change

Penis
If uncircumcised, ask patient to retract foreskin Smegma is a cheesy, white discharge and is normal under foreskin
Note location and describe any masses, ulcers, sores, or rashes
Penile discharge is always abnormal, record color, consistency and amount. A smear and culture may be ordered

Testes
Ask patient to hold penis out of way during exam
Both testes should be visible in scrotal sac
Palpate testes simultaneously, should be equal in size and shape
Normal testes are round, smooth, without nodules or masses and freely movable

The spermatic cord can be palpated from the epididymis to the inguinal ring

Female Physical Examination

Inspection And Palpation:

Breast:

Patient should be bare to the waist in sitting position with arms at sides. Inspect for size, symmetry, dimpling. Breast size may vary from right to left breast slightly. Difference in nipple direction may indicate mass

Redness may indicate infection or inflammatory carcinoma. A prominent venous pattern in one breast may indicate carcinoma

Have patient raise arms above head and repeat inspection. Have patient place hands on hips and repeat inspection Position changes may bring out dimpling/flattening in breast

Note any discharge from breast and type

Palpate breast in supine, have patient raise one arm over head while you examine other breast. Compress all of breast to include axilla in systematic manner. If nodules note location, size, shape, consistency, and mobility

External genitalia:

Separate the labia and inspect and palpate for nodules, sores, ulcers, rashes, or discharge. Note location if present

Internal genitalia

If examination is to be performed, gather equipment and assist as needed:

Slides
Lubricant
Light
Speculum of correct size
Fix or cotton tip applicators
Cervical scrape

MALE REPRODUCTIVE SYSTEM PHYSICAL ASSESSMENT FORM

Inspection:

Breast for symmetry_____dimpling_____discharge____ulcers____________________

Genitalia for level of sexual maturity______________________________________

Pubis: pediculosis pubis__________rash_________hernia______________________

Penis: circumcised_________discharge___________ulcers/sores__________________

Nodules________rash_________inflammation________other______________________

Testes:descended______other____

Palpation:

Breast: enlargement______discharge__________masses_________________________

Pubis: masses___________________pain on palpation___________________________

Penis: foreskin retraction_______hygiene____________masses___________________

Tenderness___________________other____

Scrotum: nodules_______ulcers______tenderness_______masses_________________

Spermatic cord: palpable________intact_________course_______________________

Testes: size_________shape________tenderness_____masses____________________

Assessment notes:

____

____

____

____

____

____

FEMALE REPRODUCTIVE SYSTEM PHYSICAL ASSESSMENT FORM

Inspection:

Breast: sexual maturity of breasts________________symmetry__________________

Shape_____________dimpling___________nipple discharge_________ulcers___________

External genitalia: sexual maturity level____

Pediculosis pubis______________nodules_________ulcers__________sores_____________

Rash_____________________inflammation_________________other_________________

Internal genitalia: cervix: color____________nodules_____________masses__________

Ulcerations____________discharge____________position__________other_____________

Vagina: mucosa color________________inflammation_____________ulcers___________

Palpation:

Breasts: mobility____________________consistency___________tenderness___________

Lumps/masses: location________________size_________________mobility____________

Nipple discharge___________________________axillary nodes_____________________

Internal bimanual examination: tenderness________________nodules______________

Cervix for position______________shape_________mobility_________tenderness______

Ovaries: size______________shape__________mobility__________tenderness__________

Assessment notes:

____

____

____

____

____

____

Section III – LABORATORY & DIAGNOSTIC TESTS

The following tables contain common laboratory and diagnostic tests.

Test Name	Indications	Comments
HCG - Female Human chorionic gonadotropin Positive during pregnancy peaks at 8 weeks	Pregnancy Missed menstruation Positive also in hydatidiform mole choriocarcinoma	**Regarding Collection:** First urine of day required Test 14 days after missed period Dilute urine may not be accurate Home testing kits available OTC Laboratory testing can also be done on blood serum
Estrogen - Female Early menstrual 50-300 pg/ml Mid-menstrual 100-600 pg/ml Late-menstrual 80-450 pg/ml	Timing ovulation Gonadal hypofunction Some tumors	**Regarding Collection:** Use red-top tube collect 5-10ml Indicate phase of menstrual cycle **Results:** Decreased in ovarian failure, menopause, anorexia, or Turner's Increased in some tumors
FSH - Follicle Stimulating Hormone Preovulation and postovulation 6-50mUU/24 hours Postmenopause Greater 50mUU/24hr	Infertility Menstrual problems Turners syndrome Precocious puberty Ovarian tumors	**Regarding Collection:** Collect 24-hour urine sample Obtain container from laboratory No food or drink restrictions **Results:** Decreased level tumors of testes, ovaries, adrenal gland or in anorexia nervosa Increased level in menopause, pituitary tumor or Klinefelter's
Nuteinizing Hormone (LH) Preovulation and postovulation 5-25 mlIU/ml Midcycle 30-90mIU Postmenopause Greater 35	Menstrual problems Infertility Turner's syndrome Postmenopause	**Regarding Collection:** Blood serum required Tests usually ordered with FSH **Results:** Increased in ovarian failure postmenopause, Turner's or in polycystic ovary syndrome

Progesterone **Preovulation** 20-150 ng/dl Midcycle 300-2400 ng/dl	Determine ovulation Determine corpus luteum function	**Regarding Collection:** Blood serum required **Results:** Increased during ovulation, five days post-ovulation and pregnancy
Pregnanediol **Preovulation** 0.5-1.5 mg/24hour Midcycle 2-7 mg/24 hours Postmenopause 0.1-1.0 mg/24 hr.	Menstrual problems Threatened abortion Tumor of ovary Tumor of breast Ovarian cyst	**Regarding Collection:** Collect 24 hour urine sample Obtain container from laboratory Record date of LMP on label **Results:** Decreased menstrual problems, threatened abortions, tumor
Prolactin - Female **Nonlactating** 0-23 ng/dl Pregnancy 10-20 times of normal	Pituitary tumor	**Regarding Collection:** Blood serum required **Results:** Increased pregnancy, pituitary adenoma, hypothyroidism, lack of menstruation

Section IV -- PROCEDURES AND CONDITIONS

Methods Of Contraception

Contraception (birth control) is used to avoid pregnancy. There are a variety of methods, the more common are listed.

Table of Contraception Methods

Method	Pregnancy Rate* Contraindications	Side effects	Comments
Abstinence	0% None	No risk of STD** None	Only method with 100% effectiveness
Surgical Contraception	Less than 1% Future desire for children	Rare: infection	Usually permanent Several types Tubal ligation Vasectomy
Oral Contraception	0.3% - 2% Age over 40 Chronic illness Hypertension Liver problems Stroke Heart disease	See specific agent Nausea headaches Heart attack Thromboembolism Stroke, migraines	Synthetic hormone Large variety Positive side effects: Regular periods; less cramping Major problems are rare Prescription required
Condom and Foam	1-2% Foam alone 4-29% Condom alone 2-10%	Mild vaginal irritation	Condom should be used only once Available OTC Protection from STDs
Sponge	5% Menstrual flow	Rare: Rash Irritation Toxic Shock Syndrome (TSS)	Soft polyurethane sponge in vagina Contains spermaticidal Available OTC Leave in place 6 hours after intercourse
Diaphragm	10-18% alone 3-5% with spermaticidal jelly Uterine prolapse Cystocele	Rare: TSS	Circular rubber device inserted into vagina Needs to be fitted Leave in place 8 hours after intercourse

Basal & Billings Body Temperature	6-10% None	None	Instruction by physician or nurse practitioner required to assess fertile periods Abstinence is used during fertile periods
Withdrawal	15-25% None	None	Penis is withdrawn prior to ejaculation Unreliable Not recommended
Douching after sex	35-40% None	None	Unreliable Not recommended

*Pregnancy rate over 1 year period
** STD = sexually transmitted diseases

Remember:

- Most of these forms of birth control are available at family planning clinics for low or no cost
- Only condoms, and some spermaticidal agents are effective against sexually-transmitted diseases for individuals active with more than one partner. Oral contraceptives, IUD's, diaphragms are not effective
- The onset of puberty in girls is 9-17 years
- The onset of puberty in boys is 10-19 years
- Pregnancy can occur the first time a woman has intercourse
- Pregnancy can occur during menstruation
- Pregnancy can occur up to two years after the last menses in menopausal women (documented cases even after this)
- Surgical contraception does not alter sexual desire

Sexually Transmitted Diseases (STDs)

STDs are diseases spread by sexual contact or intercourse with an infected individual

Some general comments about STDs are:

- An STD can be contracted at first sexual intercourse
- Pregnancy does not provide immunity from STDs
- A woman can become pregnant and acquire an STD at the same time
- A vaginal or penile discharge IS NOT present in all STDs
- Oral contraceptives are not effective in preventing STDs
- STDs have not proven to be transmitted by toilet seats
- Multiple partners increase the risk of contracting STDs

- Sterility can be a complication from untreated STDs
- Condoms are effective in preventing the spread of STDs
- Immunity is not developed in most STDs, reinfection is possible
- No vaccinations are available to prevent venereal disease

Symptoms indicating need for STD workup:

MALES	FEMALES
Penile discharge: white, clear, green	Vaginal discharge: irritating, yellowish, malodorous, frothy, cheesy, gray, causes intense itching
Rectal discharge	Rectal discharge
Burning on urination	Burning on urination
Frequent urination	Frequent urination
Chancres or sores on penis, testes	Chancres or sores on vulva, vagina
Sores may be painless	Sores may be painless
Itching in genital area	Itching in genital area
	Lower abdominal pain (PID)

SEXUALLY TRANSMITTED DISEASES

Disease	**Gonorrhea "Clap, GC"**	**Syphilis "Syph"**
Cause	Bacterium Nesisseria gonorrhea	Bacterium Treponema palladium
Symptoms	May be asymptomatic especially in women Penile thick yellow to green discharge 3-9 days post contact Lower abdominal pain Painful urination Fever/chills may occur Rare - arthritis, endocarditis or meningitis	Three stages of disease: #1 Painless chancre sore on genitals or mouth 10-30 days post contact No discharge noted Chancre disappears in a few weeks by itself #2 Secondary Syphilis Occurs 6 weeks + after chancre sore leaves Skin rash or lesions Fever, aches, hair loss Sore throat Clears spontaneously #3 Tertiary syphilis May appear 5-40 years after chancre sore Damages the nervous system heart and blood vessels
Treatment	Aqueous procaine Penicillin IM Ampicillin PO or Amoxicillin PO Penicillin allergies Spectinomycin IM	Benzathine penicillin IM For allergies to penicillin ErythromycinPOor PO Tetracycline hydrochloride
Complications without treatment	Urethral stricture Sterility Pelvic inflammatory disease (PID) Peritonitis	Permanent damage to major organs of the body when untreated
Infected Neonates	Gonococcal ophthalmia with blindness	Stillbirth Prematurity Congenital syphilis syphilitic lesions of organs, osteochondritis

Disease	Chlamydia	Condylomata "Venereal Warts"
Cause	Bacterium Chlamydia trachomatis	Virus Papilloma
Symptoms	Usually asymptomatic White to clear discharge from urethra or vagina Burning may occur Painful coitus Irregular menses Lower abdominal pain	Pink, cauliflower shaped warts on genitals Warts appear 1-3 months after contact Increased in pregnancy
Treatment	Tetracycline PO Erythromycin PO	Cauterization of warts Cryosurgery to warts Laser surgery Podophyllin Topical Warts may reoccur even with proper treatment
Complications without treatment	Male-Reiters syndrome Urethritis, arthritis Conjunctivitis Epididymitis Females - Endometritis Salpingo-oophoritis Rectal fistulas Both - Sterility	Increased growth of warts may block vagina
Infected Neonates	Stillbirth Prematurity Ophthalmia neonatorum	C-sections may be needed if warts are large
Disease	**Herpes Simplex II**	**Trichomonas**
Cause	Virus Herpes Simplex type II	Protozoan Trichomonas vaginalis
Symptoms	Painful blisters on genitalia appear 3-10 days after contact Swelling & inflammation May appear on rectum, mouth, fingers Initial blisters (vesicles) may ulcerate Reoccurrence of outbreak is common with stress Malaise or flu-like symptoms	Frothy vaginal discharge Intense itching, Burning on urination Profuse malodorous discharge Male is usually asymptomatic

Treatment	No cure Acyclovir Topical Zovirax PO	Metronidazole (Flagyl) PO or vaginally
Complications	Repeated outbreaks Possibility of fetal transmission	Intense discomfort from itching
Neonatal Infection	Congenital herpes virus infection usually occurs with ruptured membranes or vaginal contact during delivery Symptoms may include: Disseminated lesions of viseral organs leading to poor infant prognosis Central nervous system involvement, skin lesions Asymptomatic infant Caesarean section may be indicated to prevent infant infection	None

Section V -- DIETS USED DURING PREGNANCY

Nutrition During Pregnancy

Indications	**Knowledge of Pregnancy**
Comments	Weight gain is a normal expectation of pregnancy with a gain of 25-30 pounds desired. Weight gain should be distributed over the trimesters as follows: First trimester gain of 2-4 pounds Second trimester gain of 11 pounds Third trimester gain of 11 pounds
Restricted foods	No foods are restricted during pregnancy, however nutritious foods are encouraged and high calorie, low nutrition items are discouraged
Allowable foods	All of the basic food groups should be included; total calories should be 2400/day, 300 more calories per day than before pregnancy. The following should be included daily: Water -- 4-6, 8 ounce servings per day Meat group -- 2, 3-4 ounce servings per day Milk group -- 3-4 servings per day Bread/Cereal group -- 4-5 servings per day Vegetables -- 3-4 servings per day Fruits -- 3-4 servings per day Fats -- used in moderation
Other Interventions	Weight should be assessed weekly any gain over 2 pounds in one week should be reported as well as failure to gain weight. Prenatal vitamins with iron are recommended for very pregnant women, instruct women that stools may turn dark brown to black. Encourage nutritious snacks such as fruits, cheese, milk, fresh vegetables. Dieting is not recommended

Section VI – DRUGS

The tables below supply only general information, a drug handbook should be consulted prior to administering any unfamiliar drug.

ANDROGENS

Action: Naturally occurring hormones in the body (testosterone and androsterone) associated with the development of the male secondary sexual characteristics, the synthetic form is used to repress the pituitary output of FSH and LH.

Indications: Endometriosis, fibrocystic breast disease, breast cancers, hypogonadism, male eunuchism, check individual drug for purpose.

General comments: Amenorrhea is a reversible side effect and normal menstrual cycle will return within 60-90 days after treatment. Instruct patient to report any voice pitch changes as soon as possible to prevent irreversible changes. Pregnancy should be ruled out prior to treatment, weight should be monitored weekly with these medications for fluid retention.

Examples of drugs in this classification:

Generic	Trade	Comments
Danazol	Cyclomen Danocrine	Given PO Initialdosegiven during menses Used for endometriomas or fibrocystic breast disease
Fluoxymesterone	Halotestin Ora-Testryl	Given PO GI upset may occur Used in breast cancer and hypogonadism

ESTROGENS

Actions: Naturally occurring hormone in the body (estradiol, estrone) associated with the development of female secondary sexual characteristics as well as the changes produced by the female menstrual cycle. Synthetic form often used in the treatment of menopause.

Indications: Any condition where natural estrogens are decreased to include menopause, hypopituitarism and amenorrhea.

General comments: Pregnancy should be ruled out prior to use; use is also contraindicated in breast feeding women. Instruct patient to report any chest or leg pain, shortness of breath or slurred speech as this class of drugs is associated with thromboembolitic problems, as well as an increased risk of endometrial cancer. PAP smear should be performed at least yearly in all women on oral estrogens. Encourage postmenopausal women to increase vitamin D intake and exercise regularly to aid in the prevention of osteoporosis. Patient should be given package insert labeled "What you should know about estrogens" prior to initiation of therapy.

Example of drugs used in this classification:

Generic	Trade	Comment
Estradiol	Estrace	Given PO Generally taken 3 weeks, then off 1 week and cycle repeated. Administer with food to prevent/decrease nausea. Provide patient with package insert. Smoking and excessive caffeine intake should be discouraged

ORAL CONTRACEPTIVES -- COMBINATION ESTROGEN-PROGESTIN FORMS

Action: To prevent contraception by the prevention of ovulation, this is done by the repression of FSH and LH.

Indications: Birth control is the primary function of this classification of drugs.

General comments: Oral contraceptives may be obtained after physical examination and history is obtained. Contraindicated in women with history of heart disease, stroke or any thromboembolitic disorder due to increased risk myocardial infarctions, stroke, hypertension or thromboembolitic disease. Smoking should be discouraged in any women taking oral contraceptives. All women on oral contraceptives should have regular medical check-ups to assess for side effects from the medication. Oral contraceptives are available in combination forms with estrogen and progestin, or progestin only forms.

Instruct patient to take medication at the same time daily for best results. Pregnancy should be discouraged until at least 3 months after last use. Instruct patient to report these side effects immediately ie, chest pain, leg pain severe headaches, visual changes, shortness of breath. Instruct patient to report breakthrough bleeding, jaundice, unusual vaginal discharge, libido changes and hair loss. Oral contraceptives are considered to have a lower mortality rate for women under age of 40 than complications from pregnancy and child birth.

Example of drugs in this classification:

Generic	Trade	Comments
Estrogen/ Progestin Combinations	Ovcon Ovral Demulen Norlestrin *Modion Ortho-Novum Nordette	May be taken PO Should be taken at same time daily May have drug interactions, check with handbook when used concurrently with other medications

OXYTOCICS

Purpose: To promote uterine contractions, to stimulate the flow of breast milk in the postpartum mother.

Indications: Used to initiate or stimulate labor in selected term pregnancies, used for control of postpartum hemorrhage, to promote the letdown reflex in breast feeding mothers.

General comments: Cautious use is required, these preparations may lead to fetal or maternal facilitates if used incorrectly. Constant supervision of patient during use is recommended. Assess for previous uterine surgery, cephalopelvic disproportion, and fetal distress prior to use. Pulse and blood pressure should be monitored frequently (at least q 15 minutes) during use, fetal heart rate should be monitored continuously. These medications should never be used without a volume control pump, check all calculations and orders with another person. Assess for hypercontractility of the uterus in ante partum women and discontinue IV, turn client to left side, administer oxygen and notify physician.

Examples of drugs in this classification:

Generic	Trade	Comment
Ergonovine Maleate	Ergotrate Maleate	Given PO, IM, IV Used to prevent postpartum hemorrhage Check orders for route and admission time prior to delivery
Oxytocin	Pitocin Syntocinon	Given IV drip, deep IM, IV slow push or nasal spray

		Only one route should be used at any one time. Widely used for induction of labor as it stimulates muscular contraction of the uterus . Generally given IV drip with 10 units in 1 liter of IV solution started at 3 ml per hour, rate is increased by 3 ml every 20 minutes until contractions are q2-3 minutes apart and last for 40-60 seconds. Maximum rate is 120 ml/hr or 20 mU/min
Carboprost Tromethamine	Prostin/15M	Given IM Used to induce abortion in the 13-20th week of gestation Administer deep IM
Dinoprost Tromethamine	Prostaglandin F2 PGF	Given by Intraamniotic route Used to induce abortion in the 13-20th week of gestation. Monitor for pyrexia

Section VII -- GLOSSARY

Ab	Abortion
Abortion	Pregnancy termination before the fetus is viable, may occur spontaneously or be induced
Abruptio placenta	The premature separation of the placenta
Afterbirth	The placenta
Amniocentesis	The removal of amniotic fluid via the abdomen of a pregnant women to detect abnormalities or genetic problems
Amenorrhea	Lack of menses
Apgar	Assessment of color, heart rate, reflex irritability, muscle tone, and respirations of newborn new born infant; just minutes following birth
C-Section	Caesarean section
Cervix	The rounded structure located at the neck of uterus
Caesarean section	Surgical removal of fetus via an abdominal incision
Circumcision	Surgical removal of the prepuce of the penis
Coitus	Sexual intercourse
Condom	Device effective in the prevention of pregnancy and venereal disease, and consists of a thin flexible sheath that fits over the penis to collect semen
Contraception	Method to prevent pregnancy
Cryptorchidism	Undesended testes
D & C	Dilation & Curettage

Dilation & Curettage	A surgical procedure where the cervix is dilated and inner wall of the uterus is scraped
Dysmenorrhea	Difficult or painful menstruation
Dyspareunia	Difficult or painful intercourse for women
Ectopic pregnancy	Pregnancy where implantation occurs outside the uterus; may be in the fallopian tube, abdomen or ovary
EDC	Estimated date of Confinement, expected delivery date
Embryo	Product of contraception from 2nd to 8th week of gestation
Endometritis	Inflammation of the uterine lining
Estrogen	Hormone produced by the female ovary needed for the development of secondary sexual characteristics and menstrual cycle
Fetus	Product of contraception from the 8th to 40th week of gestation
FSH	Follicle Stimulating Hormone -- hormone secreted by the anterior pituitary gland responsible for the maturation of the follicle (oocyte and surrounding cells) prior to ovulation in the monthly menstrual cycle
G	Gravida means pregnancy usually stands for total number of pregnancies Gravida I is one pregnancy, Gravida V is five pregnancies.
GC	Gonorrhea
GU	Genitourinary -- referring to the genitals and urinary organs
Genitalia	The organs of reproduction

Gravida	Pregnancy, the number of pregnancies
Gynecomastia	Enlargement of the male breast tissue
HCG	Stands for Human Chorionic Gonadotropin
Hysterectomy	Removal of the uterus
Human Chorionic Gonadotropin	HCG -- a hormone secreted by cells surrounding a developing embryo. A positive HCG is indicative of pregnancy
LH	Luteinizing Hormone -- hormone secreted by the anterior pituitary just prior to ovulation
LMP	Last menstrual period
Leukorrhea	White discharge from the vagina
MP	Menstrual period
Mammogram	A study of the breast tissue for the early detection and cancer
Mastectomy	Removal of breast tissue, usually when cancer is present
Menarche	The onset of menstruation
Menorrhea	Normal menses or normal menstrual flow
Menopause	The cessation of menses
Neonate	Newborn
OB	Obstetrics
P	Para, number of viable pregnancies
PAP smear	Papanicolaou smear, a screening test for the detection of cervical cancer
Para	Number of viable pregnancies

Pediculosis pubis	"Crabs," a lice infestation of the hair in the genital region
PID	Pelvic inflammatory disease
Placenta previa	The placenta is implanted low in the uterus often called "low lying placenta", placenta may completely cover the cervix or just be implanted low
Primagravida	Woman pregnant for first time
SGA	Small for Gestational Age
Sterility	Inability to conceive
Testosterone	One of the male sex hormones produced by the testicles, it aids in the development of secondary sexual characteristics, is needed for libido and erection of the penis
Tubal ligation	Surgical method of contraception where the fallopian tubes are blocked to prevent fertilization
Venereal infection/disease	Sexually-transmitted infections that include gonorrhea, syphilis, chancroid, trichomonas, genital herpes, and venereal warts.

Endocrine System

ENDOCRINE SYSTEM

Table of contents

Section I -- OVERVIEW

Primary functions

1. Endocrine system synthesizes and releases hormones into the bloodstream: hormones which originate in an organ or gland are then transported to another site (via bloodstream) where they are modulators of body (e.g. increased function) and cellular responses (produce secretions) that induce a generalized or local effect on control/regulation of:

Water and electrolyte balance/metabolism
Gastrointestinal function
Growth
Metabolism and energy
Reproductive function
Stress
Inflammatory process
Behavior

2. Controls release of catecholamines, epinephrine and norepinephrine, which are chemical mediators synthesized by nerve cells and released from nerve endings in the adrenal medulla and central nervous system.

3. Production of hormonal peptides such as endorphins and enkephlins involved with pain and emotion.

Components and function

The endocrine system is involved in the control and integration of many body functions via the secretion of hormones to include the hypothalamus, pituitary gland, thyroid gland, parathyroid glands, adrenal glands, the pancreas, gastrointestinal tract, heart and kidney.

Hypothalamus and the pituitary gland (hypophysis)

The hypothalamus receives input from all areas of the central nervous system and synthesizes the following neurohormones which either stimulate or inhibit pituitary hormone release.

1. Thyrotropin releasing hormone (TRH) -- Stimulates the production and release of thyroid stimulating hormone (TSH) and prolactin (PRL) by the anterior lobe of the pituitary gland.

TSH needed for synthesis by the thyroid of T4 and T3 used in metabolism along with estrogen and progesterone stimulates breast development and milk production during pregnancy.

2. Gonadotropin releasing hormone (GnRH) -- Also called luteinizing-hormone releasing hormone (LHRH) which stimulates synthesis and secretion of luteinizing hormone (LH) and follicle-stimulating hormone (FSH) from the anterior lobe of the pituitary gland.

LH and FSH are involved with development of ovarian follicles, spermatogenesis, secretion of estrogens, ovulation and the development of the corpus luteum.

3. Dopamine (also called Prolaster Inhibiting Hormone) -- Inhibits the release of PRL, can also inhibit the release of LH, FSH and TSH under certain circumstances.

4. Corticotropin releasing hormone (CRH) -- Stimulates the synthesis and release of adrenocorticotropic hormone (ACTH) from the anterior lobe of the pituitary gland.

ACTH stimulates the adrenal cortex to secrete cortisol and aldosterone as well as some weak androgens.

5. Growth hormone-releasing hormone (GRH) -- Stimulates growth hormone (GH) release from the anterior lobe of pituitary gland.

Disorders associated with GH include acromegaly, dwarfism and gigantism.

6. Somatostatin -- Inhibits the synthesis and secretion of GH and TSH hormones from the anterior lobe of the pituitary gland.

7. Pro-opiomelanocorticotropin (POMC) -- Found in the hypothalamus, and anterior and posterior lobes of the pituitary gland. This hormone gives rise to ACTH, B-lipotropin, B-melanocyte stimulating hormone, enkephalins, and endorphins.

Enkephalins and endorphins are morphine-like peptides that bind and activate opioid receptors throughout the CNS.

The following hormones are also produced in the posterior lobe of the pituitary gland:

> 1. Antidiuretic hormone (ADH) or vasopressin -- Acts on the collecting tubules in the kidneys to prevent excessive loss of water and also promotes contraction of smooth muscle around blood vessels
> 2. Oxytocin -- Acts on the smooth muscle of the uterus to produce contractions

Thyroid gland

Located immediately below the larynx in the middle portion of the neck this gland synthesizes these thyroid hormones.

1. Thyroxine (T4) is the major hormone secreted and T4 serum levels are commonly used to measure thyroid function.

2. Triiodothyronine (T3) is secreted in smaller amounts than T4, however, its effect is more potent than T4.

Thyroid hormones are secreted in response to thyroid-stimulating hormone (TSH) produced in the anterior pituitary gland and thyroid releasing hormone (TRH) from the hypothalamus.

Thyroid hormones have several major effects:

Increase metabolism (affecting all major organs, except brain, gonads and spleen)

Necessary for physical and mental growth and development in children

Necessary for the attainment of sexual maturity

Endocrine disorders related to this gland include hyperthyroidism and hypothyroidism.

Parathyroid glands

Located on the lower edge of the thyroid gland, these glands secrete parathyroid hormone (PTH) which regulates calcium-phosphorus metabolism (requires vitamin D).

Adrenal glands

The adrenal glands are located on the apex of each kidney. The adrenal cortex secretes three major types of hormones glucocorticoids, mineralocorticoids, and adrenal sex hormones. The adrenal medulla secretes epinephrine and norepinephrine which are sympathomimetic.

1. Glucocorticoids -- Active in protecting against stress and maintaining normal metabolism, hydrocortisone or cortisol is the most abundant; corticosterone is another, Cortisone is a circulating glucocorticoid but is a metabolic by-product of cortisol.

2. Mineralocorticoids -- Help maintain water and electrolyte (sodium and potassium) balance; aldosterone is the primary mineralocorticoid.

3. Sex hormones -- Secretes both male (androgens) and female (estrogens) gonadocorticoids or sex hormones in small amounts. Estrogens are also produced by the ovaries and placenta. Testosterone (androgen) is produced by the testes.

4. Epinephrine and norepinephrine -- Involved in the "flight or fight" response; increased blood pressure, heart rate, respirations, and decreased blood to the digestive tract during times of stress.

Pancreas

Digestive accessory gland that secretes the following hormones:

1. Insulin -- Secreted by beta cells of the islets of Langerhans; this hormone is essential for metabolism. It provides for glucose storage, prevents fat breakdown and increases protein synthesis.

The only hormone known to have a direct effect in lowering blood sugar by facilitating its transport into skeletal muscle and adipose tissue.

2. Glucagon -- Secreted by alpha cells in the islets. Stimulates the breakdown of glycogen and fats and the release of glucose by the liver.

Opposite effects of insulin

3. Somatostatin -- Secreted by delta cells in the islets. Inhibits the secretion of insulin, glucagon and growth hormone.

Section II -- ASSESSMENT

Assessment of the Endocrine System

Assessment of the endocrine system involves looking at the body in general for signs of endocrine failure.

Health history

Chief complaint

Chief complaints can be widely varied depending upon which hormones are affected, therefore, complaints are grouped according by:

Growth hormone excess

Acromegaly
- Coarse facial features
- Thick skin and nails
- Wide hands and feet
- Weakness
- Impotence, infertility
- Diplopia
- Joint deformities
- Joint pain
- Deep voice
- Diaphoresis
- Hirsutism

Growth hormone deficiency

Simmond's disease
- Pallor, dry skin
- Weight loss, emaciation
- Recurrent infections
- Lethargy, decreased strength
- Impotence, decreased libido
- Intolerance to cold temperature
- Hypotension
- Atrophy of gonads and thyroid
- Amenorrhea
- Decreased perspiration
- Decreased pubic hair

Thyroid-excess:
- Hyperthyroidism
- Anxiety, mood swings
- Heat intolerance
- Tachycardia, palpitations
- Exophthalmos
- Diarrhea
- Weight loss
- Diaphoresis
- Increased hunger

Thyroid-deficiency:
- Hypothyroidism
- Mental sluggishness, fatigue
- Cold intolerance, hypothermia
- Edema
- Alopecia, coarse hair
- Constipation
- Weight gain
- Decreased diaphoresis
- Anorexia

Hyperparathyroidism:
- Renal colic or calculi
- Dysrhythmias
- Constipation, obstruction
- Anorexia, weight loss
- Nausea and vomiting
- Depression, mental dullness
- Fatigue, mood swings

Hypoparathyroidism:
- Convulsions
- Dysrhythmias
- Abdominal spasms
- Dyspnea, laryngeal stridor
- Positive Trousseau's sign
- Lethargy, personality changes
- Visual disturbances

Osteoporosis, deep bone pain
Pathologic fractures
Muscle weakness

Tingling of fingers
Calcification of ocular lens
Muscle spasms

Adrenal-hypercortisolism

Cushing's syndrome

Weight gain
Hirsutism
Amenorrhea
Weakness and fatigue
Pain in joints
Ecchymosis
Edema
Hypertension
Purple striae on abdomen
Buffalo hump
Moon face
Poor wound healing
Recurrent infections
Muscle wasting

Adrenal-hypocortisolism

Addison's disease

Weight loss, anorexia
Decreased pubic hair
Depression
Weakness and lethargy
Hypoglycemia
Bronzed skin pigmentation
Dehydration, thirst
Orthostatic hypotension
Diarrhea
Nausea

Diabetes mellitus

Weight loss, anorexia
Polydipsia, polyphagia
Acetone breath
Weakness, fatigue
Dehydration, polyuria
Increased thirst
Frequent infections
Poor wound healing
Retinopathy, blurred vision
Sexual dysfunction
Impaired kidney function-kidney failure

Personal and family history

Thyroid/Pituitary/Diabetes/Cushing's/Addison's

Hyperthyroidism, goiter, diabetes, and growth disorders may be genetic in origin.

Weight changes

Increased weight -- Hypothyroidism and Cushing's syndrome

Decreased weight -- Addison's, growth hormone deficiency, hyperthyroidism, hyperparathyroidism, and diabetes mellitus

Mood changes

Depression, decreased self-esteem, changes in body image, and changes in sexual desire or performance all affect mood and often occur with endocrine system disorders

Bowel pattern changes

Constipation -- Hypothyroidism and hyperparathyroidism

Diarrhea -- Addison's and hyperthyroidism

Frequent infections/poor wound healing

Poor wound healing and frequent infections are noted in diabetes, growth hormone deficiency, and Cushing's

Excessive thirst, hunger and urination

Classic signs of diabetes

Temperature intolerance

Heat intolerance -- Hyperthyroidism

Cold intolerance -- Growth hormone deficiency and hypothyroidism

Changes in hair distribution

Decreased pubic and axillary hair -- Growth hormone decrease and Addison's

Hirsutism (excess growth of hair or hair in unusual places) -- Growth hormone excess and Cushing's Alopecia -- hypothyroidism

Endocrine system testing

Testing will be done based on assessment findings and may include general tests such as CBC, platelets and SMA series as well as specific hormone testing, scans or biopsy

Current information

Physician -- List primary physician

Medications -- List all medications taken: Assess for medications that may effect hormone synthesis or release

ENDOCRINE SYSTEM ASSESSMENT FORM

Chief complaint

Patient's statement__________ Onset__________ Symptoms__________

Frequency__________ Duration__________ Other areas affected__________

Have you had this before?__________ Date__________

What have you done for this?__________

What do you think caused this to happen?__________

What changes have you had to make in your life?__________

Personal and Family History

	Patient Date	Family Member		Patient Only
Thyroid problems	______	______	Recent weight change	______
Pituitary problems	______	______	Fatigue/weakness	______
Diabetes	______	______	Mood changes	______
Cushing's disease	______	______	Eating habit change	______
Addison's disease	______	______	Bowel pattern change	______
Growth disorders	______	______	Frequent infections	______
Parathyroid disorder	______	______	Poor wound healing	______

Temperature intolerance__________ Change in hair distribution__________

Excessive: Thirst__________ Hunger__________ Urination__________

Endocrine system testing

Cortisol__________ Sodium__________ Calcium__________ Phosphorus__________ Ketones__________

ACTH stimulation__________ T3__________ T4__________ Protein-bound iodine__________

17-ketosteroids__________ 17-hydroxycorticosteriods__________ Glucose__________

Glucose__________ Glucose tolerance testing__________ Basal metabolic rate__________

Biopsy__________ Scans__________ Ultrasonography__________ Other__________

Current history

Physician__________ Phone__________

Medications

Name__________ Dose__________ Frequency__________ Route__________

Name__________ Dose__________ Frequency__________ Route__________

ENDOCRINE SYSTEM ASSESSMENT FORM

Inspection:

General appearance____________________________Mood____________________

Features____

Visual acuity____________________________Retinal changes____________________

Skin color________________Skin turgor______________Hair distribution__________

Wounds/infected areas____

Lesions/rashes on skin____

Nutritional status____________________________Fat distribution__________________

Palpation:

Temperature________________Pulse________Respirations__________B/P______/______

Skin texture abnormalities______________________Reflex testing____________________

Response to pinprick____________________Proprioception_________________________

Thyroid___________________________Muscle mass_______________________________

Auscultation

B/P: lying________/________sitting________/________standing________/________

Assessment Notes

____

____

____

Physical Assessment

Physical assessment of the endocrine system is done in the following order:

1. Inspection
2. Palpation

The primary task of this examination is to assess the body for signs of endocrine system failure. Patient should be undressed, gowned and covered with a sheet.

Inspection

General appearance/Mood

Appearance and mood may or may not be affected depending upon the endocrine disorder and the severity of the problem.

Descriptive terms for appearance may include: anorexic, frail, healthy, masculine, obese, physically fit, robust, stuporous, well-nourished.

Descriptive terms for mood may include attentive, anxious, cooperative, depressed, drowsy, emotionally labile, excited, inattentive, lethargic, mentally dull or slow.

Features

Coarse features can occur with growth hormone excess.

Skin

Skin characteristics -- Color (bronze, normal, pale, yellow), temperature (cool, hot, warm), texture (diaphoretic, dry, elastic, leathery, moist, thick), turgor (good, fair, poor).

Hair distribution -- Hirsutism (abnormal hairiness) or sparse hair distribution especially in pubic or axillary regions.

Lesions -- Assess color, location, size, temperature.

Diabetic skin abnormalities -- Diabetics are more susceptible to infection and usually have decreased microcirculation. Both of these problems can lead to skin disorders to include:

Bullosis diabeticorum -- Formation of a large lesion that is filled with fluid generally on the forearm, feet, fingers or toes.

Diabetic dermopathy -- Multiple hyperpigmented areas on the legs ranging in size from 0.5-2 cm in circular or oval shape.

Eruptive xanthomas -- Firm, yellow 4-6 mm lesions with a red base that appear suddenly on elbows, knees, buttocks or any site of trauma.

Necrobiosis lipoidica diabeticorum -- Red to red-brown plaques that may be yellow in the center, accompanied by shiny transparent skin, usually found on the shin; lesions may ulcerate.

Necrolytic migratory erythema -- Bright red patches found on the lower abdomen, groin, buttocks and thighs, blisters are present and quickly break leaving crusts.

Palpation

Vital signs

Temperature elevation -- Frequent infections may occur with several types of endocrine disorders to include growth hormone deficiency, diabetes, and Cushing's syndrome.

Pulse -- Rapid pulse may occur with hyperthyroidism.

Blood pressure -- Hypotension may occur with growth hormone deficiency and Addison's disease. Hypertension may occur with Cushing's syndrome.

Skin

Scleredema -- Thickening of the skin due to deposits in the dermis which may not be visible.

Section III -- LABORATORY AND DIAGNOSTIC TESTS

The following tables contain the more common laboratory and diagnostic tests used for this system.

Table of Common Laboratory Tests

Test Name	Indications	Comments
Acetone/ketone bodies Acetone 0.3-2.0 mg/dl 51.6-344.0 umol/L Ketones 2-4 mg/dl	Uncontrolled diabetes Diabetic ketoacidosis Starvation Malnutrition Excessive vomiting Excessive diarrhea	**Regarding collection:** Use red-top tube collect 3-5 ml **Results:** Increased in ketoacidosis and starvation
ACTH (Corticotropin) 80 pg/ml when measured 8am-10am	Cushing's syndrome Cancer of adrenal gland Steroid use Pituitary cancer	**Regarding collection:** Use green-top tube collect 15 ml Send on ice to laboratory **Results:** Decreased in Cushing's, adrenal gland cancer and steroid use Increased in pituitary cancer
Aldosterone (serum) 6-25 ug/24 hours	Adrenal cortical malfunction Diabetes mellitus Cancer of adrenal gland	**Regarding collection:** Patient must be supine for 1 hour before collection Use red-top tube collect 5 ml Write the time on specimen (peaks in am decreased in pm) **Results:** Decreased in Cushing's, diabetes stress and overhydration Increased in dehydration, cancer of adrenal gland, and adrenal cortical hypofunction
Cortisol 5-23 ug/dl 138-635 nmol/L 8-10 am 3-13 ug/dl 83-359 nmol/L	Cushing's syndrome Addison's disease Thyroid malfunction Diabetic acidosis	**Regarding collection:** Use green-top tube collect 15 ml Send on ice to laboratory **Results:** Decreased in Cushing's, cancer of the adrenal gland and steroids Increased in adrenal hypofunction and pituitary neoplasm

Glucose (FBS) Fasting blood sugar 70-110 mg/dl **Glucose (OGTT)** Oral glucose tolerance test 1/2 hr. <160 mg/dl 1 hour <170 mg/dl 2 hour <125 mg/dl 3 hour 70-110mg/dl **Glucose 2 hr. Postprandial (serum)** less than 140 mg/dl two hrs. after high CHO meal	Hypoglycemia Diabetes mellitus Pancreatic cancer Cushing's syndrome Burns, injuries Surgery, MI Gestational diabetes mellitus	**Regarding collection (FBS)** Use gray-top tube collect 5-10 ml NPO for 12 hours before testing Insulin is given if ordered **Regarding collection (OGTT)** High CHO diet for 3 days before NPO for 12 hours before testing Collect FBS, if high consult with physician, OGTT is usually not done if FBS is over 200 mg/dl Collect urine specimen Give 50-100 grams of glucose Use gray-top tube collect 5-10 ml at 1/2, 1, 2 and 3 hours, collect urine specimens at same times **Regarding collection (Postprandial)** Food is restricted 8 hours before High CHO meal at breakfast or lunch Use gray-top tube collect 10 ml two hours after meal is finished **Results:** Increased in diabetes mellitus, Cushing's, stress and injury Decreased in hypoglycemia, adrenal gland hypofunction and hyperinsulinism
Pregnanetriol (urine) 0.4-2.4 mg/24 hour males 0.5-2.0 mg/24 hour females	Adrenal hyperplasia Pituitary problems Adrenal tumor	**Regarding collection:** 24 hour urine collection Keep urine refrigerated Label with collection time and name **Results:** Decreased in anterior pituitary hypofunction Increased in adrenal hyperplasia, and adrenal gland tumor
(TSH) Thyroid stimulating hormone 2-5.4 uIU/ml <10 uU/ml	Hypothyroidism Pituitary problems Klinefelter's Thyroiditis	**Regarding collection:** Use red-top tube collect 5 ml **Results:** Decreased in hypothyroidism with pituitary problems Increased in primary hypothyroidism thyroiditis
Thyroxine (T4) 4.5-11.5 ug/dl serum thyroxine T4 by column	Thyroid disorders Pituitary problems Myasthenia gravis	**Regarding collection:** Use red-top tube collect 5 ml Radioisotopic substances are contraindicated

1.0-2.3 ng/dl serum thyroxine Free T4		**Results:** Decreased in hypothyroidism and anterior pituitary hypofunction Increased in hyperthyroidism, thyroiditis and myasthenia gravis
Triiodothyronine (T3) 80-200 ng/dl	Hyperthyroidism Thyrotoxicosis	**Regarding collection:** Use red-top tube collect 5-10 ml List drugs that may interfere with thyroid function **Results:** Decreased in trauma and severe illness Increased in thyrotoxicosis, hyper-thyroidism
17-Ketosteriods (17-KS) 8-25 mg/24 hours males 5-15 mg/24 hours females	Adrenal dysfunction Pituitary problems	**Regarding collection:** Usually requested with 17-OHCS 24 hour urine collection Use preservative or keep on ice Label with age, sex, name and time **Results:** Decreased in hypopituitarism, myxedema and Addison's Increased in ACTH therapy, Cushing's adrenal cancer, hyperpituitarism
17-Hydroxycor-ticosteriods (17-OHCS) 5-15 mg/24 hours males 3-31 mg/24 hours females	Addison's Hypopituitarism Hypothyroidism Cushing's Extreme stress Adrenal cancer Hyperpituitarism Hyperthyroidism	**Regarding collection:** 24 hour urine collection Preservative or refrigeration is required List any drugs taken on lab slip Coffee is restricted **Results:** Decreased in Addison's, hypopituitarism, hypothyroidism Increased in Cushing's, stress, adrenal cancer, hyperpituitarism

Table of Common Diagnostic Tests

Test Name	Indications	Comments
Computerized tomography (CT) Normal adrenal gland size	Adrenal problems	**Pre-procedure:** Explain procedure to patient Signed consent form may be needed NPO for 3-4 hours before scan

		Remove jewelry and metal objects Assess for iodine allergy Takes 30-60 minutes **Post-procedure:** No activity restrictions
Radioactive iodine (RAI) uptake test 2 hours 1-13% 6 hours 2-25% 24 hours 15-45%	Thyroid tumor Thyroid malfunction	**Pre-procedure:** Assess for any allergy to iodine or seafood Test contraindicated in pregnancy NPO for 8 hours before testing Explain procedure to patient Radioiodine capsule or radioiodine liquid will be given and x-ray studies will be done at regular intervals (2,6, and 24 hours) Patient may eat one hour after the radioiodine is ingested All medications are to listed on laboratory slip Inform patient take radioiodine will not harm visitors **Post-procedure:** Assess for hyperthyroidism, report if present
Thyroid Scan Normal size, structure and position of thyroid	Thyroid malfunction	**Pre-procedure:** Explain procedure to patient Instruct patient that for 7-10 days thyroid and cough medications are discontinued and iodine-rich foods are restricted Signed consent form is required A radioactive isotope will be given IV during the test **Post-procedure:** Assess for reaction to isotope
Ultrasonography Normal structure and size of thyroid, adrenals, or parathyroid depending on area scanned	Thyroid malfunction Adrenal malfunction Parathyroid malfunction	**Pre-procedure:** Procedure is non-invasive Takes 30-60 minutes A lubricant will be applied and transducer moved over area **Post-procedure:** No activity restrictions

Section IV -- PROCEDURES AND CONDITIONS

Diabetes Mellitus

A disorder of carbohydrate metabolism with a total or relative lack of the hormone insulin and elevated blood glucose levels. It can lead to blindness, kidney failure, limb amputation and death.

Classic symptoms

Polyuria from diuretic effect of elevated blood glucose, with glucosuria and hyperglycemia, thirst, hunger and weight loss all occur.

Types of diabetes

Insulin-dependent diabetes mellitus (IDDM) or Type I

- Dependent on insulin injections to sustain health and maintain life
- Accounts for 5-10% of known cases of diabetes
- Have frequent, major fluctuations in blood glucose levels
- Genetic component linked to MHC antigens (HLA) located on chromosome 6 (found in 90% of white Type I diabetics); autoimmunity and viruses have also been implicated
- Family history of diabetes is minor; 10% have a parent or sibling with DM
- Onset in infancy, childhood, or young adulthood usually before age 30
- Often a lean body build is present
- Usually requires a combination of intermediate-acting (NPH) and short acting (regular) insulin for control
- Ketoacidosis will develop without insulin injections
- Continuous subcutaneous insulin infusion with pumps or three or more daily insulin injections may be required
- Requires frequent blood glucose monitoring
- Self-monitoring of blood glucose (SMBG) is usually done at home Results are available quickly and the procedure can be done frequently
- Diet therapy is of major importance to maintain consistent intake of food and to synchronize food intake with insulin injection as well as exercise

Non-insulin-dependent diabetes mellitus (NIDDM) or Type II

- May or may not use insulin for control of symptoms
- Accounts for approximately 85-90% of all cases of diabetes
- Blood glucose profile is more stable in Type II than in Type I

- Family history is marked for diabetes
- Onset is usually after age 30, disorder develops slowly over weeks, months or years
- Includes obese (60-90% of all IDDM cases) and non-obese NIDDM
- Variable insulin production with insulin resistance
- If treated with oral agent or insulin, the risk of hypoglycemia is present
- Primary dietary goal is to achieve and maintain ideal weight; this may control symptoms without medications

Gestational diabetes mellitus

- Affects 2-3% of pregnant women, therefore, the American Diabetes Association recommends that all pregnant women undergo a 1 hour oral glucose challenge test with 50 grams glucose
- If glucose challenge test is 140 mg/dl or greater, then oral glucose tolerance testing (OGTT) is recommended
- Gestational DM is diagnosed if more than two of the following values are met or exceeded:
  1 hour OGTT of 190 mg/dl or greater
  2 hour OGTT of 165 mg/dl or greater
  3 hour OGTT of 145 mg/dl or greater
- Often onset is in 2nd-3rd trimester
- Infants born to mothers with gestational DM are at risk for larger infants (macrosomia), delayed fetal lung maturation, neonatal hypoglycemia and even fetal demise
- Fetal well-being is monitored via fetal activity determinations (diary of fetal movement is kept), contraction stress testing, and fetal ultrasounds to rule out macrosomia
- Not an indication for cesarean section
- May disappear at the end of the pregnancy
- Risk factor for the subsequent development of diabetes, therefore, post-pregnancy testing should be done periodically

Complications of diabetes

Acute complications

Hyperglycemia

- Relative or absolute lack of insulin with excess circulating stress hormones leads to increased blood glucose
- Requires exogenous insulin if absolute for correction, relative lack may be treated with diet alone, oral hypoglycemic agents, or exogenous insulin

Hypoglycemia

- Most common complication of insulin treatment in Type I DM
- Estimated that Type I diabetics experience one episode of hypoglycemia per week
- Can occur when insulin taken is excessive, when a meal is delayed, has insufficient carbohydrate intake, or gastric emptying time is delayed
- Insulin excess can occur with exercise or change of insulin injection site
- Alcohol intake, onset of menses, can also lead to hypoglycemia
- Severe hypoglycemia may result in seizures or coma

Diabetic ketoacidosis (DKA)

- Affects 2-5% of Type I (insulin dependent) diabetics per year
- Hyperglycemia and ketosis
- Nausea and vomiting related to ketoacidosis
- Sepsis occurs in 1-10% of patients with up to 40% having an infection
- See table on diabetic ketoacidosis for signs, symptoms, lab values, and management

Diabetic Ketoacidosis (DKA)

Signs and symptoms	Lab values	Management
Nausea and vomiting	Glucose q 1-2 hours	Admit to hospital ICU lifethreatening
Abdominal pain very severe	200-2000 mg/dl	
	600 DKA average	Give insulin
Dehydration	70-110-normal	10 U Regular as bolus IV
Poor skin turgor		10 U Reg/hour via IV
Dry mucous membranes	Ketones (plasma)	
	1:2-1:64	Unresponsive after 4 hours 2-10 fold increase (Reg)
Hot flushed skin	1:16 DKA average	
Respiratory distress	Negative-normal	
Tachypnea		Corrected acidosis decrease to 1-2 U Reg/hour
Kussmaul respirations	HCO-3 q 4 hours	
	4-15 meq/L	
Shock	10 DKA average	Fluids for replacement
Hypotension	24-28-normal	0.9% saline, 2-3 liters within first 3 hours, K+ given at 10-30 meq/hour
Tachycardia		
Fever -- possible infection	Blood pH PRN	
	6.80-7.30	Then 0.45% saline at 150-300 ml per hour until blood glucose is 250 mg/dl then 5% glucose is added and infusion slowed
Mental status	7.15 DKA average	
Confusion	7.35-7.45-normal	

Progressive LOC Coma General Possible local infection Monitor vital signs and mental status every 1-2 hours Continous ECG monitoring	Pco2 14-30 mmHg 20 DKA average 35-45 normal Frequent monitoring of all lab values during treatment to assess progress	Monitor weigh q 6-12 hours Monitor ECG Flattened/inverted T waves and U waves indicate hypokalemia Tall T waves, widened QRS and loss of P waves with hyperkalemia Monitor output Assess every 1-2 hours Foley catheter as needed

Hypoglycemia

Signs and symptoms	Lab values	Management
Faintness, weakness Tremulousness Palpitations Diaphoresis, cold skin Hunger, nervousness Confusion Increased pulse rate	Glucose less than 50mg/dl	Oral glucose ingestion 2-3 tbsp sugar in water or fruit juice (orange juice) If unable to swallow give 0.5-1 mg Glucagon SQ or IV If no response after 20 min. give IV glucose Give oral carbohydrates when patient is alert

Hyperglycemic-hyperosmolar nonketotic coma

Signs and symptoms	Lab values	Management
Same as DKA without nausea and vomiting plus Neurological Generalized seizures Focal seizures Reversible hemiparesis Dehydration More severe than DKA Severe hypotension Very important that neurological status is monitored frequently	Glucose q 1-2 hours 600-2000 mg/dl 1000 mg average Ketones (plasma) not significant Osmolality 360 mosM average 280- 300-normal Blood urea nitrogen 65 mg/dl average Creatinine 3.0 mg/dl 0.6-1.2 - normal	Admit to hospital ICU, life-threatening problem Fluids for replacement 0.9% saline, 1 L in 30 min. If hypotension persists 1 L in 30-60 min. 0.45% sodium chloride in at 500 ml/hr, 2-3 liters then infusion is slowed Fluid replacement continues until positive balance is achieved, may take 24-36 hours When blood glucose is 250-300 mg/dl add 5% glucose Monitor I&O q 1-2 hrs and weight q 6-12 hrs Insulin therapy as in DKA

Hyperglycemic hyperosmolar coma (HHC)

- Usually occurs in adults over 50 years of age with Type II (non-insulin dependent) diabetes
- In 35% of HHC cases diabetes has not yet been diagnosed
- Onset is slower than in DKA and medical attention is usually obtained later
- Ketosis is absent
- Conditions that may precipitate HHC include MI, pancreatitis, sepsis, or stroke
- Dehydration is more severe than in DKA, dehydration is due to diminished kidney function
- May present with seizures, myoclonic jerking, or hemiparesis
- Mortality ranges from 12-42%

Infections

- Glucosuria is associated with an increased in vaginitis
- Minor trauma to tissues can lead to infection secondary to vascular insufficiency and decreased sensory neuropathy
- Infections include cellulitis, soft tissue necrosis, draining wounds
- Diabetic patients are susceptible to severe infections such as gangrene, mortality with gangrene is greater than 10%
- Malignant otitis externa is almost exclusive for diabetics, pain and drainage from the external canal are present, infection can reach the cranial nerves, meninges or sigmoid sinus, death may occur from epidural abscess
- Urinary tract infection is common in diabetics due to high urine glucose concentration

Chronic complications (usually occur after 10 years)

Microvascular disease

- Capillaries are thickened
- Renal disease or failure if the glomerular capillaries are effected, 50% of diabetics with IDDM have renal failure after 20-30 years
- Retinal disease occurs with visual loss if retinal capillaries are affected, 50% of diabetics have some degree of problems after 10 years

Macrovascular

- Diabetics have an increased incidence of large vessel disease such as atherosclerosis
- Risk of death from cardiovascular disease is 3.5 times higher than that of non-diabetic
- Hypertension and diabetes is present in 2.5 million Americans

- 30% of diabetics develop peripheral vascular disease
- Diabetics require 5 times more amputations than non-diabetics

Neuropathic diseases

- Demyelination and degeneration of the nerves may occur with clinical neuropathy present
- Nerve involvement may lead to pain, sensory loss, motor weakness, or loss of position sense deep tendon reflex loss

Treatment of diabetes

Diet

- The goals of nutrition management is to improve blood glucose, to obtain consistent nutrient intake on a day to day basis, to obtain and maintain desired weight, to promote healthy eating habits individualized dietary intervention is vital for compliance

Medications

- Insulin is required for IDDM
- Sulfonylurea drugs such as tolbutamide, tolazamide, glipizide, and glyburie may be used in NIDDM
- Metformin a guanidine derivative is also used to treat NIDDM
- Glucosidase inhibitors are used in the treatment of hyperglycemia

Section V -- DIETS

Diets used in diabetes mellitus

Food exchange list

Indications	IDDM, NIDDM, or gestational diabetes One of several meal-planning approaches that may be used in diabetes
Comments	Goal is to improve blood glucose and lipid levels; diet is consistent in day-to-day intake Calories are sufficient to achieve and maintain desired weight Cholesterol is limited to less than 300 mg/day Sodium is allowed up to 3000 mg/day unless a restriction is needed Fiber up to 40 grams/day unless calories are restricted then 25 grams/1000 calories Alcohol is limited to 1-2 equivalents per week Diet is individualized and planned by dietitian Carbohydrates should make up 55-60% of diet Proteins should make up 12-20% of the diet Fat should make up less than 30% of intake Polyunsaturated fats, up to 10% Saturated fats, less than 10% Monounsaturated fats 10-15% Exchange plan is made up of six lists of core foods which have approximately the same distribution of carbohydrates, proteins and fats within each list

Exchange list	Carbohydrate grams (g) X4	Protein g X 4	Fat g X 9	Calories
Breads/ Starch	15	3	trace	80
Meats				
Lean	0	7	3	55
Medium fat	0	7	5	75
High fat	0	7	8	100
Vegetables	5	0	0	60
Fruits	15	0	0	60
Milk				
Skim	12	8	trace	90
Low fat	12	8	5	120
Whole	12	8	8	150
Fats	0	0	5	45

DIABETIC EXCHANGE LIST

LIST 1
Starches & Breads

1/2 cup cold cereal
- Raisin Bran®
- 40% bran flakes

1/3 cup cold cereal
- All Bran®
- Bran buds

3/4 cup cold cereal
- Cheerios®
- Cornflakes®
- Grape Nut Flakes®
- Kix®
- Product 19®
- Rice Krispies®
- Wheaties®

1 1/2 cup cold cereal
- Puffed rice
- Puffed wheat

1/2 cup hot cereal
- Cream of rice
- Cream of wheat
- Oatmeal
- Wheat bran

1/2 cup pasta
- Macaroni
- Egg noodles
- Spaghetti

1/3 cup rice

1/3 cup beans
- Kidney beans
- White beans
- Lentils

LIST 2
Meats

Lean meat exchanges:

1 oz. broiled beef
- Flank steak
- Tenderloin
- Top loin

1 oz. roasted eye round beef steak

1 oz. pork
- Ham, fresh, cured, canned or broiled
- Canadian bacon
- Roasted tenderloin

1 oz. poultry roasted
- Skinless chicken, turkey or duck

1 oz. fish broiled
- Cod, haddock, halibut, salmon

2 oz. clams, steamed crab, lobster or scallops

1/4 cup tuna

1/4 cup cottage cheese

1 oz. lite cheese

3 egg whites

1/2 cup egg beaters

LIST 3
Vegetables

1/2 cup cooked:
- Asparagus
- Beets
- Broccoli
- Brussels sprouts
- Cabbage
- Carrots
- Cauliflower
- Chinese cabbage
- Greens
- Green beans
- Green peppers
- Leeks
- Mushrooms
- Okra
- Onions
- Peapods
- Sauerkraut
- Spinach
- Summer squash
- Tomatoes
- Turnips
- Zucchini

1 cup raw
- Bean sprouts
- Broccoli
- Carrots
- Cauliflower
- Onions
- Peppers
- Tomatoes

1/2 cup juice
- Tomato
- Vegetable

LIST 1
Starches & Breads (cont.)

Black-eyed peas
Split peas

1/2 cup vegetables
Corn, whole
Corn, creamed
Lima beans
Garden peas
Mashed potatoes

3 oz. baked potato

1 sliced bread
White
Whole wheat
Rye
Pumpernickel
French

1/2 pt., hamburger or frankfurter bun

2 bread sticks -- 1/2 by 4 1/2 inch

1 six inch corn or wheat tortilla

3 graham crackers, 2 1/2 inch square
8 animal crackers
5 melba toast
24 oyster crackers
3 cups plain popcorn
6 saltines

LIST 2
Meats (cont.)

Medium fat exchanges:
1 oz. beef
Ground beef - regular to extra lean
Cubed steak
Chuck pot roast
Porterhouse steak
Rib roast
Rump roast
T-bone steak
Meatloaf

1 oz. pork
Center loin roast
Center loin chop
Pork cutlet
Top loin chop

1 oz. veal cutlet

1 oz. poultry, roasted
Chicken, turkey, duck or Goose with skin

1 oz. ground turkey

1/4 cup tuna canned in oil

1 oz. cheese
American, skim
Mozzarella
Weight Watcher's®

1 egg
1/4 cup egg substitute

4 oz. tofu

LIST 3
Vegetables (cont.)

Free foods

1 cup:
Cabbage
Celery
Cucumber
Lettuce
Mushrooms
Radishes
Spinach

LIST 1 Starches & Breads (cont.)	LIST 2 Meats (cont.)	LIST 3 Vegetables (cont.)
	1 oz. liver, heart, or kidney	
	High fat meats 1 oz. corned beef or prime rib	
	1 oz. ground pork or spareribs	
	1 oz. fried fish	
	1 oz. cheese American, blue, cheddar, Monterey or Swiss	
	1 oz. luncheon meat, smoked sausages, brat-wurst, or frankfurters	
	1 tbl. peanut butter	

LIST 4 Fruits	LIST 5 Mild	LIST 6 Fats
1 two-2 1/2 inch Apple Nectarine Orange Peach	**Skim/very low fat** 1 cup of: Skim milk 1/2% milk 1% milk Buttermilk low-fat	**Unsaturated fats** 1 teaspoon of: Margarine, soft/hard Mayonnaise Corn oil Cotton seed oil Olive oil Peanut oil Safflower oil Soybean oil
1/2 cup of: Applesauce Cherries Peaches* Pears* Apple juice	1/3 cup non-fat dry milk 8 oz. plain non-fat yogurt	

LIST 4
Fruits (cont.)

Orange juice
Pineapple juice

3/4 cup of:
Blackberries, raw
Blueberries, raw
Grapefruit*
Mandarin oranges*
Peach, sliced
Pineapple

1/2 or raw fruit
Banana
Grapefruit
Mango, small
Papaya
Raspberries

1 1/4 cup of
Strawberries
Watermelon

Free foods
1/2 cup of:
Cranberries, whole/raw
Rhubarb

LIST 5
Mild (cont.)

Low-fat milk
1 cup 2% milk
8 oz. low fat yogurt

Whole milk
1 cup whole milk
1/2 cup evaporated milk

8 oz. yogurt, plain

LIST 6
Fats (cont.)

1 tbl. of:
Margarine, reduced calorie
Mayonnaise, reduced calorie

2 tbl. of reduced calorie salad dressing

1/8 avocado
10 olives
6 almonds
10 large peanuts
2 pecans or walnuts

Saturated fats
1 slice bacon

1 tsp. of:
Butter
Shortening

2 tbl. of:
Coconut
Light cream
Sour cream

1 tbl. of :
Heavy whipping cream
Cream cheese

Free foods:
Bouillon, 8 oz.,
Soda*, 12 oz.
Club soda, 10 oz.
Coffee, 6 oz.
Taco sauce, 3 tbl.
Gelatin*, 4 oz
Equal, 1 package
Tea, 6 oz.
Dill pickle
Hard candy*
Mustard, 1 tsp.
Jam/jelly*, 2 tsp.
Pancake syrup*, 2 tbl
Whipped topping, 2 tbl.
Catsup or horse-radish, 1 tbl.

* Unsweetened or sugar free

Sample exchange plans

	1300kcal	1500kcal	1800kcal	1800kcal
Breakfast	2 bread 1 fruit 1/2 milk* 1 fat	2 bread 1 fruit 1/2 mild* 1 fate	3 breads 1 fruit 1 milk* 1 fat	3 breads 1 meat** 1 fruit 1 milk* 1 fat
Lunch	2 bread 1 meat** 1 vegetable 1 fat	2 bread 2 meats** 1 vegetable 1 milk* 1 fruit 1 fat	2 breads 2 meats** 1 vegetable 1 fruit	2 breads 2 meats** 1 vegetable 1 fruit 1 milk* 1 fat
Snack	1 fruit	1 bread	1 bread	1 fruit
Dinner	1 bread 3 meats** 2 vegetables 1 milk* 1 fat	1 bread 3 meats** 2 vegetables 1 fruit 1 fat	2 bread 3 meats** 2 vegetables 1 fruit 2 fats	2 bread 3 meats** 2 vegetables 1 fruit 1 fat
Snack	1 bread 1/2 milk*	1 bread 1/2 milk* 1 fat	1 bread 1 fruit 1 fat	1 bread 1 fruit 1 fat

* Based on low fat milk list

** Based on medium fat meat list

Section VI -- DRUGS

Drugs Used In Endocrine Disorders

This table should be used for general guidelines, a drug handbook should be consulted for more in-depth information.

INSULIN

Action: Decreases blood glucose

Indications: Absolute lack of insulin production or insulin production that is insufficient to maintain safe levels of blood glucose, used with a dietary program.

General comments: Insulins are available in short-acting (regular-rapid onset), intermediate-acting (NPH or lente) and long-acting (ultralente or protamine zinc insulin).

Insulin may be derived from the pancreas of a cow (bovine), or pig (porcine), or be commercially produced by recombinant DNA technology. Enzymatic conversion of pork insulin to a product that has the same amino acid sequence as human insulin, and therefore, is better tolerated than pork (P) or beef (B) insulins.

Insulin therapy attempts to mimic the natural pattern of insulin secretion and meet the needs of the individual patient some possible insulin therapies include combinations of the three types.

Regular insulin SC at least 20-30 minutes before each meal to mimic the natural increase in insulin secretion after a meal and an Intermediate-acting insulin at bedtime and in a small morning dose, these insulins peak in 8-10 hours after injection so bedtime dose should peak at breakfast, morning dose provides additional daytime coverage.

Long-acting insulin given in 1 or 2 injections daily, may also be combined with regular insulin before meals, long acting peaks in 12-16 hours after injection and sustains its action up to 24 hours

Insulin therapy is individualized and determined by response to the plan. Target blood sugar (BS) levels are used to determine effectiveness; target BS is also individualized. The following is a possible target BS in a healthy young person with IDDM.

Before meals BS should be between 70-130
1 hour after meal BS should be 100-180
2 hours after meal BS should be 80-150
Pre-dawn BS (0200-0400) should be between 70-120

Before administration check expiration date, do not use if outdated, discolored, thick, or clumpy.

A major adverse effect of insulin is the risk of hypoglycemia. Instruct patient is may include sweating, nausea, weakness, confusion, tachycardia, and double vision.

Examples of drugs in this classification:

Generic	Trade	Comments
Short acting Regular insulin	Velosulin (H) Iletin II (B) Iletin II (P)	Onset 15 minutes Duration 5-8 hours Only insulin that can be given IV Regular insulin should be clear
Intermediate insulin	Insulated NPH (H) NPH Iletin II (B) NPH Iletin II (P)	Onset 2 hours Duration 18-24 hours Modified insulin with addition of NPH (neutral protamine hagedorn) Usually mixed with regular insulin and given twice daily, also available pre-mixed with 30% regular insulin and 70% NPH Appears cloudy or milky Always mix by gentle rotation between palms
Premixed insulins	Humulin 70/30 Mixtard 70/30	Peaks 2-12 hours after injection Combination NPH and Regular Onset is 30 minutes

SULFONYLUREA DRUGS

Action: To augment insulin secretion, thereby, lowering blood glucose levels; not for use in IDDM where no natural insulin is secreted.

Indications: NIDDM where insulin secretion is decreased or insulin action is impaired.

General comments: Not an oral insulin, all insulin must be injected. Sulfonylurea drugs are prescribed when diet and exercise alone have not been effective in controlling blood sugar.

Compliance with medication regime is important in preventing hyperglycemia and hypoglycemia; alcohol use is not recommended with this medication.

Examples of drugs in this classification:

Generic	Trade	Comments
Acetohexamide	Dymelor	Given PO in single or divided dose May have diuretic action Assess for dehydration
Chlorpropamide	Diabinese Glucumide	Given PO with breakfast Assess for skin rashes, jaundice or easy bleeding/bruising and report Peaks in 2-4 hours Long duration
Glipizide	Glucotrol	Given PO, 30 min. before breakfast May be given twice daily Peak level in 1-3 hours Short duration of effect 4-8 hrs.
Glyburide	Diabeta Micronase	Given PO, 15-30 min. before breakfast May be prescribed in single dose or divided dose (give before dinner) Peak effect in about 4 hours
Tolbutamide	Orinase	Given PO, may be given once daily or in divided doses to prevent GI upset Peaks in 3-4 hours Duration is 6-10 hours
Tolazamide	Tolinase	Given PO with breakfast Absorbed slowly Peaks in 3-4 hours Duration up to 20 hours

ANTITHYROID AGENTS

Action: Interferes with the synthesis of T4 and T3.

Indications: Hyperthyroidism, iodine-induced thyrotoxicosis.

General comments: Reversal of hyperthyroid state is goal with weight gain and decreased pulse noted with treatment. Treatment may be long-term, close monitoring is required, patient should report any skin rash or sore throat.

Assess for signs of hypoprothrombinemia to include ecchymoses, purpura, petechiae, or bleeding. OTC medications should be avoided.

Examples of drugs in this classification:

Generic	Trade	Comments
Methimazole	Tapazole	Given PO at same time daily as food may alter response 10 times more potent than PTU Does not induce hypothyroidism
Potassium iodide	Losat KI	Given PO after meals or IV IV in thyroid crisis
	Pima SSKI Thyro-block	PO med should be given with full glass of water or juice not milk Assess for GI bleeding, pain or vomiting Sudden withdrawal may lead to thyroid storm
Propylthiouracil	Propyl-tyracil PTU	Given PO at the same time daily as food may alter response Discontinue 3-4 hrs. before radioactive iodine treatment

THYROID AGENTS

Action: Used as replacement or substitution therapy

Indications: Hypothyroidism

General comments: Frequent laboratory monitoring of thyroid levels is needed to determine clinical response, early responses to therapy include diuresis, increased pulse rate, increased appetite, loss of constipation.

Instruct patient to take exactly as prescribed, therapy is life-long. Patient may want to discontinue therapy when symptoms of hypothyroidism subside.

Trade names of medications are not interchangeable, hormone content may vary.

Examples of drugs in this classification:

Generic	Trade	Comments
Levothyroxine	Synthroid Eltroxin Levothroid Noroxine Synthrox	Given PO before breakfast or IV in myxedema coma or stupor, IV medication reconstituted with NaCl just prior to administration Shake vial until clear

Liothyronine sodium	Cytomel Tertoxin	Given PO Used in T3 suppression test
Liotrix	Euthyroid Thyrolar	Given PO before breakfast Report headache
Thyroglobulin	Proloid	Given PO, tablet may be crushed and mixed with fluids or food Dosage is adjusted by protein-bound iodine at 4-8 mcg/dl
Thyroid	Armour thyroid Thyro-teric Westthroid	Given PO on an empty stomach Single dose Avoid foods high in iodine to include seafood, turnips, cabbage soybeans, and some breads

Section VII -- GLOSSARY

Acromegaly	Anterior pituitary hypersecretion of growth hormone with enlargement of the hands, feet and face
Adrenalectomy	Surgical removal of adrenal glands
Adrenalitis	Inflammation of the adrenal glands
Aldosteronism	Adrenocortical hyperfunction with muscular weakness, excessive thirst and urination
Basal metabolic rate	Thyroid function test to measure the rate of energy expenditure or use
Cretinism	Infantile hyperthyroidism, symptoms include growth and mental retardation
Diabetes insipidus	Metabolic disorder with polyuria and polydipsia, caused by insufficient amounts of antidiuretic hormone
Dwarfism	Growth disorder caused by anterior pituitary growth hormone hypofunction or hyposecrelion
Euthyroid	Normal function of thyroid gland
Exophthalmos	Eyes that are protruding
Giantism	Excessive growth beginning in adolescence caused by excessive secretion of growth hormone
Hirsutism	Abnormal hairiness
Hypophysectomy	Removal of the pituitary gland
Myxedema	Adult hypothyroidism
Parathyroidectomy	Surgical removal of parathyroid glands
Polydipsia	Excessive thirst

Polyuria	Excessive urination
Radioisotope	Isotope that is radioactive; used as a tracer to diagnose thyroid function
Thyrasthenia	Weakness caused by hypothyroidism
Thyroidectomy	Surgical removal of thyroid gland
Thyroid crisis	Exacerbation of existing hyperthyroidism; may be caused by trauma, surgery, or severe adrenocortical insufficiency
Thyroiditis	Inflammation of the thyroid gland
Thyrotoxicosis	Thyroid crisis

Musculoskeletal System

MUSCULOSKELETAL SYSTEM

Table of contents

Section I -- OVERVIEW

Primary functions

- To provide the supporting framework for the body
- To provide internal support and protection to tissues and organs
- To enable body movement and postural changes
- To serve as storage depots for calcium, magnesium and phosphorus
- The manufacture of blood cells (hemopoeisis) within the bone marrow

Components and physiology:

The musculoskeletal system is composed of the bones, cartilages, muscles, joints, ligaments and tendons of the body

Bones (Skeletal System)

The bones combine with the muscles to provide body support, protection of the internal organs, and allow for mobility. In addition, bones store calcium, magnesium, and phosphorus. Red bone marrow manufactures red blood cells (erythropoiesis) and white blood cells.

The skeletal system contains 206 bones and can be divided into two sub-sections, axial skeleton and the appendicular skeleton.

A. The axial skeleton includes the bones that are found in the center axis of the body. Bones found here are listed in the table below

Bones of the Axial Skeleton

Bone	Location
8 bones of the cranium	The bones that cover the brain and organs of hearing/sight
1 frontal	Forehead
2 parietal	Sides of the cranium
2 temporal	Begins at temples, continues back behind the ear
1 occipital	Posterior portion of the cranium
1 sphenoid	Bottom of cranium (butterfly shaped)
1 ethmoid	Behind the nasal bones, and above the sphenoid bone

14 facial bones	Bones of the face
2 nasal	Bridge of the nose
2 maxillae	Upper jawbone
2 zygomatic	Cheekbones
1 mandible	Lower jawbone
2 lacrimal bones	Behind the nasal bones in the eye socket
2 palatine bones	Back portion of the hard palate part of the floor/walls nasal cavity
2 inferior nasal conchae	Located within the skull behind the nose
1 vomer	Nasal septum
1 hyoid bone	This bone supports the tongue
6 auditory bones	Bones of the inner ear
2 malleus	The hammer
2 incus	The anvil
2 stapes	The stirrups
51 trunk bones	From the skull to the end of spinal column
26 vertebrae	The spinal column
7 cervical	Neck
12 thoracic	Back
5 lumbar	Lower back
1 sacral	Center bone of pelvic girdle
1 coccyx	Tail bone
24 ribs	Rib cage in front of body
14 True ribs	1-7th ribs attach to sternum
6 False ribs	8-10th ribs attach to one another
4 floating ribs	11 and 12 do not attach in front
1 sternum	Bone over the center of the chest

B. The appendicular skeleton contains the bones that are appendages to the axial skeleton including the shoulder and hip girdles, the arms and legs. The table below shows the bones found in the appendicular system and the figure on page 334

Bones of the Appendicular skeleton

Bone	Location
4 bones in the shoulder girdle	Shoulders
2 clavicles	Collarbones, just below neck in the front
2 scapulae	Shoulder blades
60 bones in upper extremities	Arms, wrist and hands
6 bones in the arms	
2 humerus	Bone between elbow and shoulder
2 radius	Bone between elbow and wrist on the thumb side
2 ulna	Bone between elbow and wrist on the little finger side
16 carpus bones	Wrist
2 navicular (Scaphoid)	
2 lunare	
2 triquetrum	
2 pisiform	
2 trapezium	
2 capitate	
2 hamate	
38 bones in the hand	
10 metacarpels	Bones in the palm
28 phalanges	Bones in the finger
2 coxal bones	Pelvic bones or hips
60 bones in the lower extremities	Legs, ankles, and feet
8 bones in the legs	
2 femurs	Thigh
2 tibia	Inside of the lower leg

The Bones of the Body

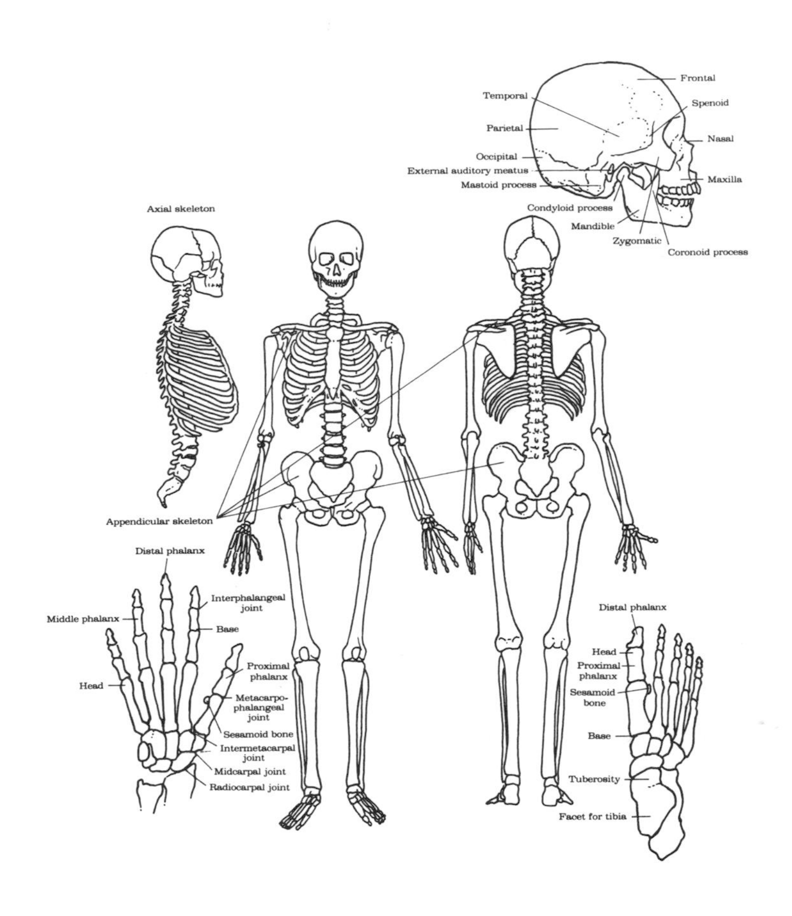

Figure 8A

2 Fibula	Outside of the lower leg
2 Patella	Kneecap
14 Tarsus Bones	Ankle
2 Talus	
2 Calcaneus	
2 Cuboid	
2 Navicular	
6 Cuneiform	
38 Bones of the feet	
10 Metatarsals	
28 Phalanges	Toes

Cartilage

Cartilage, dense connective tissue, is also an important component of the skeletal system. Cartilage is associated with the ribs, nasal septum, external ear, larynx, trachea, bronchi, vertebrae, and the articular surface of bone. Cartilage is not bone. However, it is capable of withstanding great pressure and tension. Cartilage does not have a nerve or blood vessel supply

Muscle

Muscle tissue is composed of fibers which contract or shorten. These unique elasticity properties facilitate the movement of an organ or bone. There are three kinds of muscle tissue: striated or skeletal, visceral or smooth, and cardiac

A. Striated muscle tissue attaches to bone. Skeletal muscle is striated and under voluntary control. There are more than 600 striated muscles throughout the body. It is skeletal muscle that makes position changes possible by the movement of the bones in the skeleton. Skeletal muscles are grouped by the type of movement they produce (flexors, extensors, adductors, abductors, internal rotators, external rotators, or circum flexors)

B. Visceral muscle is smooth muscle tissue found in the blood vessels, stomach, intestines, and other internal organs. Visceral muscle tissue is generally under involuntary control

C. Cardiac muscle is both striated and under involuntary control. Cardiac muscle is found only in the heart

Joints

The bones in our body are rigid and serve as levers. The joints connect the bones and allow a position change of the bone so they serve as fulcrums for the levers. The junction at which the bones and

The Muscles of the Body

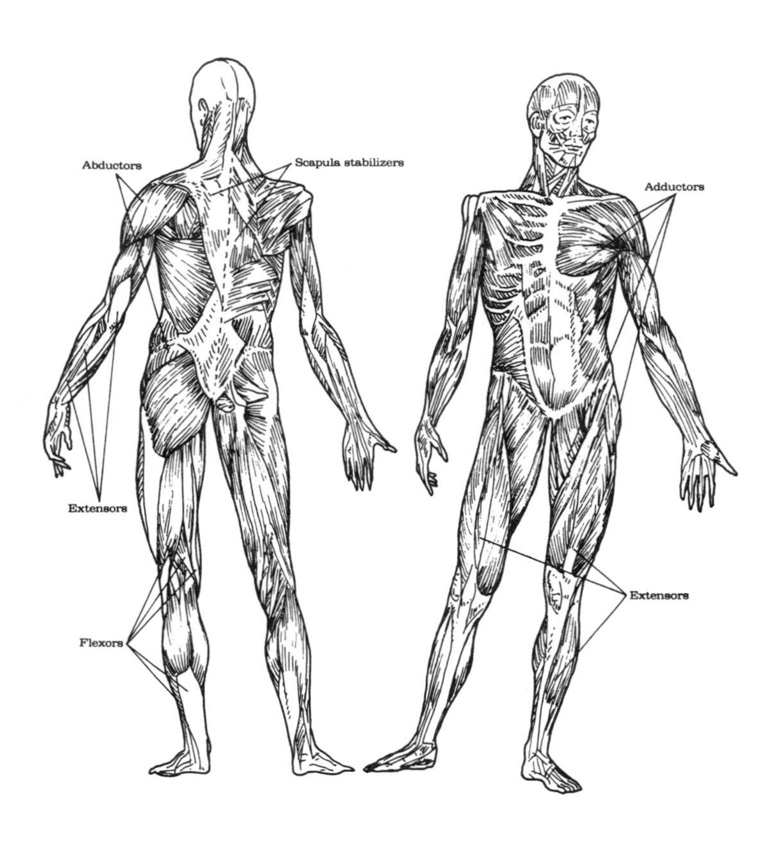

Figure 8B

joints meet are secured by ligaments or other binding tissue which can allow for much movement, or none. The joints in our body can be divided into three main types: fibrous, cartilaginous, and synovial.

A. Synarthroses joints do not move at all or only slightly. An example is found between the bones of the skull and face

B. Amphiarthroses are bones held together with a band of cartilage. An example is the symphysis pubis

C. Synovial joints are adapted for movement. Common examples include the ankle, wrist, shoulder, elbow, hip and knee. The largest joint in the body is the Tibiofemoral synovial joint (knee). The joints move body parts in seven different patterns. These are flexion, extension, abduction, adduction, internal rotation, external rotation and circumduction

Ligaments

Ligaments are bands or sheets of strong connective tissue that connect the end of the bones to create or limit movement. The ligaments are named according to location

Tendons

Tendons are bands or sheets of strong connective tissue that connect muscles to bones. One of the most familiar tendons in the body is the Achilles tendon

Section II -- ASSESSMENT

Health History

Assessment of the musculoskeletal system involves both the muscular system and the skeletal system

Chief complaint

Examples of some common chief complaints include:

Back pain or stiffness

Pain, stiffness and swelling of the joints

Sensations of weakness, numbness, tingling, or burning

Bone pain

Reduced range of motion or reduced abilities

Muscle spasms or pain

Fractures, sprains or strains

Personal and Family History

The family or personal history of the musculoskeletal disorders may supply clues that will aid in the diagnosis

Muscular dystrophy

A progressive, inherited, muscle disorder with atrophy of the muscles. The most common form is Duchenne with onset in males between ages of 3-7 years. No treatment is available

Arthritis or rheumatism

Arthritis is an inflammation of a joint and describes over 25 different diseases involving one or more joints. Rheumatoid arthritis is a syndrome with inflammation and destruction of the joints

Poliomyelitis

An acute viral infection that is highly contagious and can lead to residual paralysis. Immunization is required for all infants and children. After immunization became available the incidence of this disease is extremely low

Systemic lupus erythematosus

An inflammatory connective tissue disorder occurring mostly in women. Most patients complain of joint pain and deformity may be present. The characteristic sign is a "butterfly" rash on face

Low back pain / fractures / joint pain or swelling

Pain is a frequent, subjective complaint. For low back pain assess if an event preceded the pain such as a work injury or an automobile accident. For fractures, list year and bone fractured. Joint pain or swelling may indicate one of the arthritic disorders

Muscle pain or cramps/weakness/twitching

Muscle weakness, cramps, pain, numbness, tingling of the limbs may be related to overuse, congenital defects, or degenerative diseases. List problematic area

Musculoskeletal system testing

A variety of tests may have been completed or ordered to include x-rays, scans, biopsies, endoscopic procedures (such as arthroscopy), or blood tests. These tests include potassium, sodium, calcium, phosphorus, glucose, blood urea nitrogen (BUN), creatinine, albumin, serum glutamic-oxalacetic transaminase (SGOT), rheumatoid factor, complement fixation, lupus erythematosus cell preparation, and antinuclear antibody (ANA)

Current information

List the physician's name

List medications in current use

MUSCULOSKELETAL SYSTEM ASSESSMENT FORM

Chief Complaint

Patient's statement______________Onset_______Symptoms____________________

Frequency____________Duration___________Other areas affected______________

Have you had this before?_________________Date____________________________

What have you done for this?____

What do you think caused this to happen?__________________________________

What changes have you made in lifestyle?__________________________________

Personal and Family History

	Patient Date	Family member		Patient Only Date
Muscular dystrophy	______	______	Low back pain Back surgery	______
Arthritis or Rheumatism	______	______	Fractures and location	______
Poliomyelitis	______	______	Joint pain or swelling	______
Systemic lupus Erythematosus	______	______	Amputation	______

Muscle pain or cramps?_____________Weakness?_____________Twitching?__________

Difficulty in walking?_____________Standing?______________Sitting?____________

Combing hair?_____________Buttoning/zipping?_____________Eating?_____________

Musculoskeletal system testing

X-rays_________________Bone scans________________Muscle/bone biopsy__________

Arthroscopy_______________Arthrocetesis______________Myelogram______________

Electromyography (EMG)_________________Blood tests___________________________

Other____

Current treatments

Who is your physician?______________________________________Phone______________

Medications

Name______________________________Dose_______Frequency_______Route__________

Name______________________________Dose_______Frequency_______Route__________

Name______________________________Dose_______Frequency_______Route__________

MUSCULOSKELETAL SYSTEM PHYSICAL ASSESSMENT FORM

Inspection:

General appearance________________________Anomalies________________

Spine: Normal__________Other__________________Orthotic devices__________

Ability for self care: Roll over__________________Pull to sitting__________________

Sit supported________________sit unsupported__________transfer to chair_______

Bear weight on legs________stand____________support (type)____________________

Ambulate______________gait_______________support (type)__________________

If non-ambulatory means of locomotion____

Grasp strength________________feed self__________comb hair______dress__________

Palpation:

ROM:	Right	Left		Right	Left		
Shoulder			**Hip**			**Neck**	
Adduction	____	____	Flexion	____	____	Rotation	____
Abduction	____	____	Extension	____	____	Flexion	____
Flexion	____	____	Adduction	____	____	Extension	____
Extension	____	____	Abduction	____	____		
Internal ROM	____	____	Internal ROM	____	____		
External ROM	____	____	External ROM	____	____		
Elbow			**Knee**			**Waist**	
Flexion	____	____	Flexion	____	____	Flexion	____
Extension	____	____	Extension	____	____	Extension	____
Supination	____	____				Rotation	____
Pronation	____	____				Left	____
						Right	____
Wrist			**Ankle**				
Flexion	____	____	Plantar Flexion	____	____		
Extension	____	____	Dorsiflexion	____	____		
Radial	____	____	Inversion	____	____		
Ulnar	____	____	Eversion	____	____		
Fingers			**Toes**				
Flexion	____	____	Flexion	____	____		
Extension	____	____	Extension	____	____		

Note any asymmetry of muscle strength, joint pain, swelling, redness

Assessment Notes:

____

____

____

____

Physical Assessment

Physical assessment of the musculoskeletal system is done in the following order:

1. Inspection
2. Palpation

The assessment of this system begins by observing body positions and movements

Inspection

General appearance/Anomalies

Assess the ability to obtain, maintain, and move out of a position. Does the patient appear weak, feeble, strong. Was he able to change position easily? What body movements were observed? Was there pain on position change?

Are any contractures, amputations, or congenital anomalies present? Describe

Spine

Does the spine appear straight? Lordosis is an exaggerated lumbar curve, usually seen in obesity, pregnancy or hip deformities Kyphosis may be seen in the elderly, this is a curve of the thoracic area

Scoliosis is a lateral curve to the spine. If suspected have the patient bend forward while assessing the symmetry of the shoulder blades, hips and lateral curve in the spine

Orthotic devices

Any leg or wrist braces, twister cables or other prosthetic devices should be listed

Ability for self care:

Can the patient perform the functions listed? Do they need assistance such as an over-bed bar. Can they ambulate? Describe their gait. If they cannot ambulate, do they use a wheelchair or are they bed-bound? Assess grasp (firm, moderate, weak)? Can the patient perform fine motor skills of daily living?

Palpation

Range of motion (ROM) is tested by asking the patient perform a specific movement. Each joint should be assessed for normal or abnormal range of motion. The degree of movement (45, 90, 180) can be listed or the term full ROM or limited ROM recorded for each joint

Section III -- LABORATORY AND DIAGNOSTIC TESTS

The following tables show the common laboratory and diagnostic tests

Test of Common Diagnostic Tests

Test Name	Indications	Comments
Arthrography Knee-Normal medial meniscus Shoulder-Normal joint capsule tendon sheath intact bursa	Abnormality of the ligaments or the cartilage Ligament tears Chronic knee pain Chronic shoulder pain Performed prior to arthroscopy	**Pre-Procedure:** Check agency policy for consent Explain procedure to patient Assess for allergies to seafood or iodine Check orders for premedication such as Benadryl No food or drink restriction **Post-Procedure:** Check orders for restriction to joint use Apply ice as indicated for pain Analgesic may be given for pain
Arthroscopy Normal ligaments and tendons Normal cartilage	Joint abnormality Joint surgery Torn ligaments Arthritis Generally done on the knee	**Pre-Procedure:** Signed consent form is required Explain procedure to patient NPO if general or spinal anesthetic are given Local anesthetic may be used **Post-Procedure:** Apply ice as indicated Analgesic may be ordered for pain Assess for swelling, bleeding Rest joint for 48-72 hrs
Bone scan (nuclear scan) Normal no abnormality	Osteomyelitis Degenerative bone diseases Bone cancer Assess response to radiation therapy or chemotherapy	**Pre-Procedure:** Explain procedure to patient A radionuclide will be given IV After administration a 2-3 hour waiting period before scan Imaging will take 30-60 minutes Patient will need to lie still Signed consent may be required All jewelry must be removed Assess for iodine allergy **Post-Procedure:** No activity restrictions

Electromyography (EMG) Muscles at rest minimal electrical activity noted With contraction of muscles at work increased activity	Muscular dystrophy Myotonia Amyotrophic lateral sclerosis (ALS) Poliomyelitis Myasthenia gravis	**Pre-Procedure:** Explain procedure to patient An electrode will be inserted into selected muscles and activity will be recorded Signed consent form is required Caffeine drinks and smoking are restricted before testing (3 hrs.) Testing takes about 60 minutes SGOT, CPK, LDH if ordered should be drawn before testing **Post-Procedure:** Analgesic may be ordered
Muscle Biopsy Normal muscle No necrosis No variation in muscle fiber size	Muscular dystrophy Werdnig-Hoffman Myotonia	**Pre-Procedure:** Signed consent form is required Surgical incision to remove muscle fibers for examination is done Explain procedure to patient **Post-Procedure:** Analgesic may be ordered Assess site for drainage, or signs of infection
X-Rays Normal bones and joint structure	Arthritis Fractures Bone diseases Congenital anomalies	**Pre-Procedure:** Frequently ordered procedure Assess type of x-ray ordered Clothing and jewelry over area should be removed Testes should be shielded Assess for pregnancy in females Should be avoided in 1-3 month **Post-Procedure:** No limitations on activity

Table of Common Laboratory Tests

Test Name	Indications	Comments
Calcium (Ca++) 4.5-5.5 mEq/L 9-11 mg/dl 2.3-2.8 mmol/L	Immobility Carcinoma Multiple fracture Burns Tetany	**Regarding Collection:** Use a red-top tube collect 5-10 ml **Results:** Decreased in vitamin D deficiency, burns, infections, laxative use Increased in hypervitaminosis D, cancer, immobility. fractures

Muscle enzymes Aldolase (ALD) 22-59 mU/L at 37C	Muscular dystrophy Trichinosis	**Regarding Collection:** Use a red-top tube collect 3-5 ml **Results:** Decreased-late muscular dystrophy Increased-early and progressive muscular dystrophy, trichinosis
Muscle enzymes Alkaline phosphatase (ALP) 20-90 U/L at 30C	Liver disease Leukemia Bone cancer Healing fractures Rheumatoid arthritis Osteomalacia	**Regarding Collection:** Use a red-top tube collect 5-10 ml List medications on lab slip List age of patient on lab slip **Results:** Decreased-hypervitaminosis D, Hypothyroidism, vitamin C deficiency Increased-liver disease, cancer fractures, arthritis, osteomalacia
Muscle enzymes Creatine Phosphokinase (CPK) Male 55-120 IU/L Female 10-80 IU/L CPK-MM skeletal muscle	Muscular dystrophy Trauma	**Regarding Collection:** Use a red-top tube collect 5 ml Note number of IM injections in the past 24-48 hours **Results:** Increased total CPK/MM in muscular dystrophy, crushing injury, and in surgery Decreased in pregnancy

Section IV -- PROCEDURES AND CONDITIONS

It is estimated that more than half of all Americans exercise regularly to:

A. Increase in mental well-being
   - Improved physical appearance
   - Improved self-image and self-esteem
   - Reduction of stress and tension
   - Sleep pattern improvement
   - Increase in feelings of well-being

B. Increase in physical well-being
   - Loss of excessive weight
   - Decrease in the percentage of total body fat
   - Reduction in blood pressure
   - Increased cardiopulmonary fitness
   - Increase in muscle tone, strength and flexibility

Assessment prior to selection of exercise program:

Prior to the onset of exercise a physical assessment should be performed to determine possible risk factors to starting an exercise program. In addition to a physical a stress test may be required of:

- Males over 40 years or females over 55
- Individuals with a family history of heart disease
- Individuals with cholesterol level over 220
- Individuals with blood pressure over 160/90 mmHg
- Heavy cigarette smokers and/or heavy drinkers

For the following conditions a physical therapy, occupational therapy, or nutrition consultation may be helpful to adapt an exercise program to fit a special need and reduce the risk of injury

Intolerance to activity (walking or stair climbing)

Previous orthopedic or muscular injuries

Arthritis (limited range of motion or strength, pain)

Osteoporosis (increased risk of fractures)

Diabetes (careful monitoring of blood glucose, change in diet)

Severe obesity (more stress on weight-bearing joints)

Neurologic disorders (seizures, tremors, impaired coordination, dizziness, or impaired vision may interfere with exercise)

Guidelines for selecting an exercise program:

Must match the physical and financial capabilities present

Should be pleasurable to the individual

Should involve easily available equipment

There are two basic types of exercise programs, aerobic and anaerobic. Aerobic exercise is the most common type of program used today with the most overall benefits for the whole body. Examples of aerobic exercises include:

Aerobic dancing	Jogging	Swimming
Bicycling	Racquetball	Tennis
Canoeing	Skiing	Walking
Ice skating	Squash	

A good exercise program includes these guidelines:

- Sessions should be between three and five times a week for 20-60 minutes
- Proper body movements should be noted
- Each session has a 5-10 minute "warm up" to increase joint and muscle readiness for exercise
- After warm-up the exercise should allow for maximum use of large muscle groups in a rhythmic fashion
- The exercise should not be so strenuous that the individual is unable to talk during the activity
- The target heart rate (THR) during the activity is 60-75% of the maximum heart rate (MHR). Increase in activity intensity is determined by THR and tolerance to exercise

220 - individual's age = MHR in beats per minute (BPM)

60-75% of this number is the THR

Example.....Find THR for a 35-year-old

220	185	185
-35	X.60	X .75
185 = MHR	000	925
	1110	1295
	111.00	138.75
	low range	high range

Heart rate during exercise should be between 111 and 138 BPM

- Session should end with a "cool down" lasting 5 minutes to decrease muscle stiffness and soreness

Complications of Aerobic Exercise

Injuries or complications occur when the program is not followed correctly (no warm-up, over-doing intensity or frequency) or improper instruction has been given. Other complications include:

- Sudden death or myocardial infarction. Symptoms include chest pain, lightheadedness, dizziness, pallor, and nausea. Level of activity should be decreased if these symptoms occur
- Musculoskeletal injuries (shin splints, tendonitis, fractures)
- Joint injuries
- Inadequate hydration and/or overheating during exercise. Proper clothing for activity, avoidance of rubberized suits. Avoid exercise in hot stuffy rooms. Replacement of fluids before, during and after exercise is recommended as follows:

1-2, 8 oz glasses of cold water 15-30 minutes prior

1-3 ounces of cold water during exercise

4-8 oz of cold water after exercise

Traction

Traction is the application of a pulling force (weight) to an extremity or other part of the body. Traction is used to correct:

- Fractures
  - Immobilize fracture prior to reduction
  - Reduce the fracture
  - Immobilize fracture after reduction
  - Maintain proper skeletal alignment of the fracture
  - Realign bone fragments
  - Relieve pain by reducing muscular spasms
- Dislocated or subluxed joints
  - Maintain proper skeletal alignment
  - Prevent dislocation of subluxed joints
- Correction/prevention of skeletal deformities
  - Maintain proper skeletal alignment
- Treatment of muscular spasms/low back pain
  - Relieves pain
  - Prevents muscular spasms
- Joint replacements
  - Provides temporary immobilization of new joint
  - Maintains suture position
  - Provides skeletal alignment during early healing process
- Treatment of diseased joints
  - Allows for rest of affected joint
  - Relieves pain

Four Basic Types of Traction:

Type	Comments
Manual	Hands are used to apply force of traction Short term, used for emergency treatment of fractures Used until other traction is available Should be applied firmly and smoothly
Skin	Straps/bandages are applied to the skin Force of traction is on the patient's skin Specific amount of weight is applied Short term form of traction

	Used to prevent muscle spasms and relieve pain
	May be continuous or intermittent
	Complications:
	    Skin breakdown due to strap/bandage pressure
	    Circulatory impairment if improperly applied
	    Nerve damage

Encircling	A form of skin traction Traction device encircles extremity or body part May be continuous or intermittent Often used for cervical or pelvic traction Complications: Same as for skin traction

Skeletal	Stainless steel pins, wires, or tongs are inserted through the bone where force of traction is applied
	Continuous form of traction
	Greater weight may be applied than with other forms of traction
	Used primarily for fractures of the long bones
	Complications:
	    Infection at pin site
	    All complications of immobility

Assessment for the Patient in Traction

Assess physician orders for:

- Admitting diagnosis
- Date traction was applied or will be applied
- Type and purpose of traction to include affected body part
- Frequency of traction (Continuous/intermittent)
- Diet Ordered
- Type, frequency of physical therapy if ordered

Assess patients history for:

- Any chronic or acute illness that may affect recovery

Assess the medication record for:

- Pain medications ordered to include type, last dosage, and frequency
- Any bowel elimination medications

Assess the nurses notes for:

Patients response to traction, medications ordered for pain
Date of last bowel movement & characteristics
In take and Output record for imbalances
Last assessment findings for comparison with current condition

Assess the patient for:

Any pain or discomfort to include location
Patient's perception of traction and treatment
Patient's current mental status (depressed, anxious, relaxed)
Complaints of numbness or tingling in affected extremity/part
Proper alignment of body
Affected extremity for skin color and temperature
Blanching of nails on affected extremity
Quality of pulses on both extremities
Pin sites should be clean and dry, free of drainage, foul odor, swelling or redness (skeletal traction)
Any wrinkles in the slings or dressings
Skin for signs of breakdown, redness or discoloration
Skin for breakdown under pressure points (heels, coccyx, elbows, scapulae)
Abdomen for distention and presence of bowel sounds

Assess traction equipment for:

Correct amount and position of weight applied
Free movement of ropes through pulley and freely hanging weight
All ropes present for any frays
Assess for counter traction that prevents patient from being pulled toward the traction. Counter traction may be the persons weight or supplied by a suspended balance system

Complications of Traction:

Bone fragment motion in fractures

Can lead to severed blood vessels and/or severed nerves
Prevent with proper alignment and avoidance of contact with pin surfaces or trauma site
Assess and maintain traction frequently

Skin breakdown from pressure of traction or immobility

Can lead to decubitus ulcers
Prevent by using pressure relief mattress or sheepskin
Apply lotion via massage to pressure points
Use clean, dry, wrinkle-free linen

Circulatory impairment and/or nerve damage

Can lead to increased skin breakdown and infection
Maintain proper body alignment and assess traction equipment
Assess skin traction straps for wrinkles and correct fit
Assess for adequate circulation and report to physician if inadequate

Constipation related to Immobility

Assess bowel elimination pattern and characteristics
Increase fiber and fluids in diet while immobile
Provide stool softeners and laxatives as ordered and needed

SECTION V -- DRUGS

The tables below supply only general information; a drug handbook should be consulted prior to administering any unfamiliar drug.

ANTI-INFLAMMATORY AGENTS

Action: To decrease or prevent inflammation esp. of joints

Indications: Arthritis, to include juvenile, adult, rheumatoid, or other disorders where inflammation of the joints is a problem

General Comments: Drugs shown here vary widely, and a drug handbook should be consulted prior to administering, esp. with the gold compounds. Over the counter drug use with these medications should be checked with physician

Examples of Drugs in this Classification:

Generic	Trade	Comments
Aurothioglucose	Solganal Gold Thioglucose	Given IM Used to treat rheumatoid arthritis Familiarity with drug vital prior to administration IM injection should be deep Use gluteal site Many adverse side effects Stomatitis may occur
Gold Sodium Thiomalate	Myochrysine	Given IM Used to treat rheumatoid arthritis Same as above
Ibuprofen	Motrin Advil Nuprin	Given PO Used to treat rheumatoid arthritis, osteoarthritis musculoskeletal pain Take on empty stomach for best results Tablet may be crushed and mixed with food or drink Available over the counter
Oxyphenbutazone	Oxalid Oxybutazone	Given PO Tablet may be crushed and mixed with food Administer with food

MUSCLE RELAXANTS

Action: To decrease unwanted involuntary movement and decrease muscle tone; often used to relieve muscular pain

Indications: Muscle strains, sprains, low back pain, arthritis, bursitis, multiple sclerosis

General Comments: Decreased mental alertness may occur with use of these medications, instruct patient to avoid any hazardous activities such as driving or operating equipment until effect of medication on individual is known. Over the counter medications should not be taken without checking with physician first, instruct patient not to stop drug without medical supervision. Alcohol use should be avoided when taking these medications

Examples of Drugs used in this classification:

Generic	Trade	Comments
Balclofen	Lioresal Lioresa DS	Given PO Skeletal muscle relaxant May be taken with food
Carisoprodol	Soprodol Soma	Given PO Skeletal muscle relaxant May be taken with food
Cyclobenzaprine Hydrochloride	Flexeril	Given PO Skeletal muscle relaxant Dry mouth may be present Short-term use drug
Dantrolene Sodium	Dantrium	Given PO, IV Skeletal muscle relaxant Give IV after assessing for blood return, very irritating to tissue Assess ambulation as may be affected by this drug Monitor for jaundice during use
Tubocurarine	Tubarine	Given IV Administered by physician Onset in seconds Muscle paralysis is desired effect Used with anesthesia

SECTION VI – GLOSSARY

Abduction	Movement away from the median line of the body
Achilles tendon	Tendon located at back of heel
Active movements	Movements done by the patient without help
Adduction	Movement toward the median line of the body
Asymmetrical	The sides of the body are unequal
Amyotrophia	Atrophy of muscle tissue
Ankylosis	Immobility of a joint
Antagonist	A muscle that acts in opposition to another
Arthritis	Inflammation of one or more joints that may be chronic or acute
Arthralgia	Joint pain
Balance	Ability to maintain a steady position
Biceps	Muscle of the upper arm that flexes forearm
Closed fracture	Fracture with skin intact
Diarthroses	Joints with free movement
Diplegia	Legs are affected more than any other body part
Dorsiflexion	Foot pointing up
Eversion	Turning out
Extension	Straightening of a limb or increasing of the joint angle
Extensors	Muscles that are used to extend a joint
External rotation	Turning of a body part outward away from the midline

Flexion	Bending at a joint or decreasing the joint angle
Flexors	Muscles that are used to flex a joint
Hemiplegia	One side of the body affected
Internal rotation	Turning of a body part inward toward the midline
Inversion	Turning inward
Involuntary Muscle	Muscle that usually cannot be moved at wil
Ligaments	Bands of connective tissue that connect bone and cartilage
Muscle Tone	Degree of tension present in muscle tissue
Myalgia	Muscular pain
Myoclonus	Spasm in muscle tissue
Myokinesis	Movement of muscle
Myopathy	Disease of the muscles
Osteitis	Inflammation of the bone Osteomalacia - Softening of bone
Osteomyelitis	Inflammation of bone marrow
Osteoporosis	Bones are very porous and subject to fracture
Osteosarcoma	Tumor of bone that is malignant
Paraplegia	Lower extremities are affected
Passive movements	Movements done for the individual without his/her help
Pectoralis	Muscles in the chest
Plantar	Foot pointing down
Pronation	Facing down (palm facing down)

Spondylitis	Inflammation of the spinal column
Spincters	Circular muscles capable of constriction
Supination	Face up position (palm facing up)
Synarthroses	Joints that are immovable
Tendon	Fibrous band that connects muscle to bone
Tetany	Spasm of muscle that is paroxysmal
Tic	Muscular twitch
Triceps	Muscle in upper arm that extends forearm
Voluntary muscle	Muscle tissue moved at will

Renal System

RENAL SYSTEM

Table of contents

Section I – Overview

Primary functions

1. Controls the volume of blood

   Aids in the regulation of blood pressure by changing the amount of circulating blood volume.

2. Controls the composition of blood

   The metabolism of nutrients results in waste products within the bloodstream including carbon dioxide, water, ammonia, and urea Kidneys alter the blood removing selected amounts of various waste products and reabsorbing needed materials.

3. Aids in electrolyte and acid-base balance

   Excess sodium, chloride, sulfate, phosphate, and hydrogen ions also need to be removed from the body to maintain homeostasis.

4. Formation of urine and excretion of urine

   Excretes wastes from the body

Components

The urinary system as composed of the kidneys which produce urine, the ureters which transport urine to the bladder where urine is stored, through the urethra, and out of the body.

Kidneys

The kidneys are a pair of organs located just above the waist on the back wall of the abdomen. The kidney on the right sits slightly lower than the kidney on the left due to the location of the liver. The kidneys are about 4 inches long, 3 inches wide and 1 inch thick and are usually not palpable. Each kidney is composed of millions of functional units known as nephrons which complete the major work of the renal system. Nephrons have three primary functions:

Control blood concentration of water and solutes
Help regulate blood pH
Removal of toxic wastes from the blood

Each nephron consists of a renal tubule and a glomerulus. The glomerulus is composed of capillaries that are surrounded by a thin sac known as Bowman's capsule. Bowman's capsule exits into the tubules, first to the proximal convoluted tubule then to the Loop of Henle and finally to the distal convoluted tubule.

The Urinary System

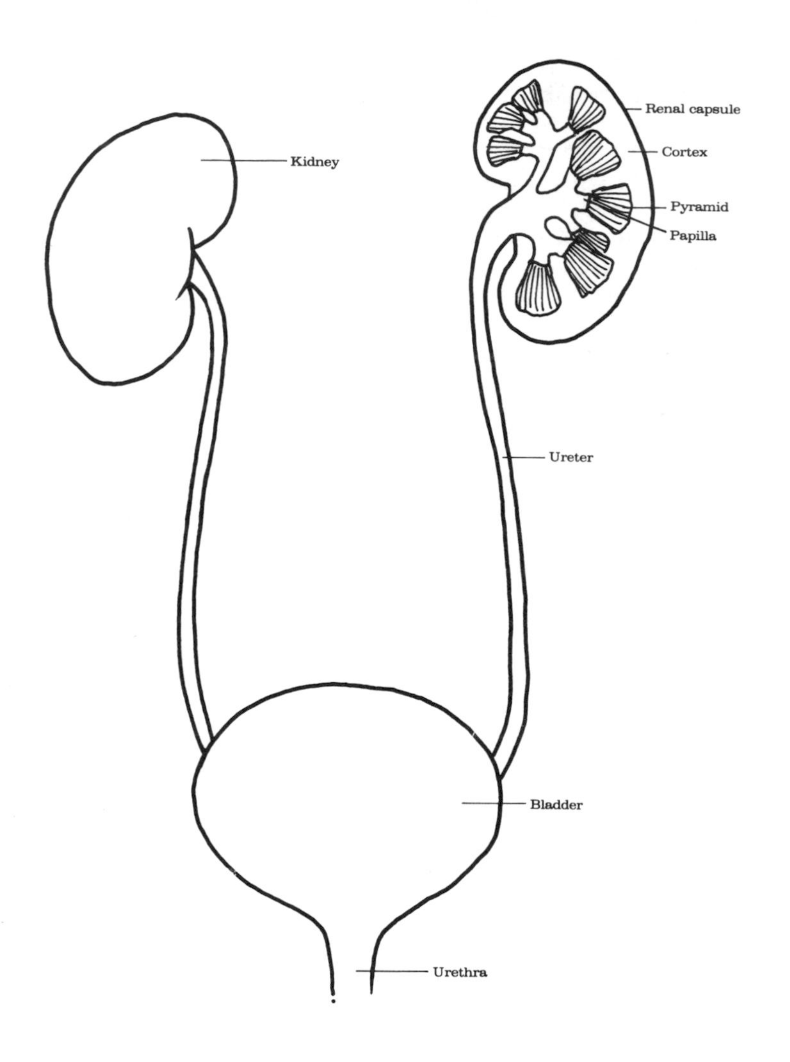

Figure 9A

Ureters

The ureters form a passageway for urine from the kidneys to the bladder. The ureters in the adult vary in length from 28-34 cm.

Urinary bladder

The urinary bladder is the storage compartment for urine

The size of the bladder varies according to how much urine it contains. The average adult bladder holds about 500 ml of urine when moderately full, although it may hold 700-800 ml. In general the urge to void is felt when the bladder contains between 200-400 ml.

Urethra

The urethra is the passageway that the urine will follow from the bladder to exit the body. In women the urethra is about 1.5 inches long; in men the urethra is about 8 inches long.

The formation of urine

The formation of urine involves three phases: glomerular filtration, tubular reabsorption, and tubular secretion.

Glomerular filtration

Blood enters the glomerular portion of the nephrons at high pressure forcing the plasma toward Bowman's capsule to be filtered through the thin membrane walls. Smaller materials such as water, glucose, amino acids, nitrogen wastes, and ions pass through the capsule and into the tubules while large materials such as the formed elements of blood and larger protein's are unable to pass. The entire volume of blood is filtered approximately 60 times a day.

Tubular reabsorption

Water, glucose, amino acids, and electrolytes such as Na+,K+, Ca, Cl, and $HC0_3$ are all reabsorbed according to body need during their passage through the nephron tubules. The specific amount of any substance reabsorbed is regulated.

An example of this regulation is the reabsorption of sodium (Na+). If Na+ concentration in the blood is low, blood pressure will drop and the renin-angiotensin pathway is activated. The release of renin by the kidney will convert angiotensinogen into angiotensin I which will be converted into angiotensin II and then angiotensin III.

Angiotensin II and III will stimulate the release of aldosterone which will cause increased reabsorption of Na+ and water into the bloodstream. This, in turn, will raise the blood pressure. Out of the 125 ml of fluid filtered every minute, most of the volume will be reabsorbed into the bloodstream; only about 1-2 ml will be eliminated as urine. Once the filtrate passes through the distal convoluted tubule, it enters the collecting tubules. The collecting tubules receive filtrate from several nephron units.

Tubular secretion

This is the final phase of urine formation. Materials such as K+ H+, ammonia, creatinine, and certain medications (penicillin) will be added to the filtrate from the blood. The purpose of tubular secretion is to eliminate waste materials and control blood pH. To raise blood pH, hydrogen (H+) and ammonium (NH4+) ions are secreted into the filtrate which makes the urine pH acidic (usually around 6).

The regulation of circulating fluid volume

If the circulating blood volume is low and the person is dehydrated, the kidneys will receive less blood to filter and less blood to supply nutrients to meet their needs. When dehydration occurs the brain releases a hormone from the posterior pituitary, known as ADH or antidiuretic hormone. ADH increases the permeability of the collecting tubules allowing for more water to return to the vascular system.

If the volume of circulating blood is increased or if no ADH is released, the collecting tubules are less permeable and less water is able to move back to the vascular system. More water is then excreted, and the urine is diluted in appearance.

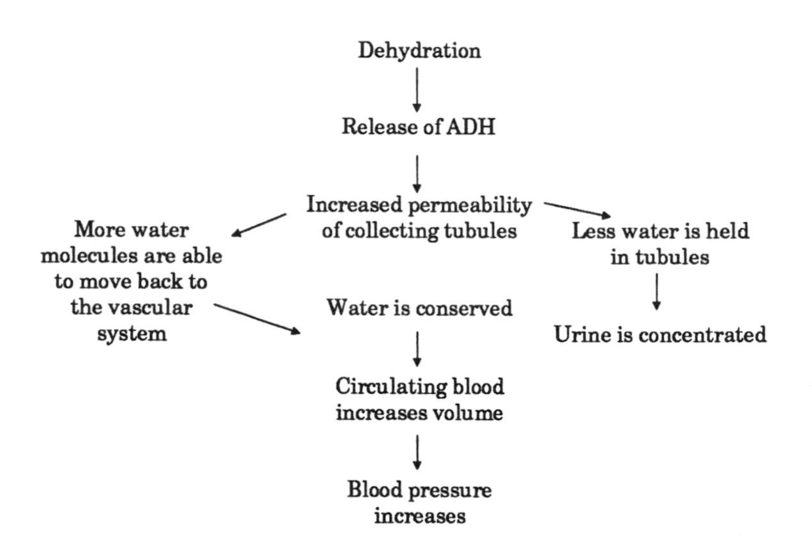

Section II -- ASSESSMENT

Some individuals may find it difficult to discuss problems of the urinary system such as incontinence and a supportive, non-judgmental manner is important.

Health history

Chief complaint

Examples of common chief complaints include:

Increased urinary frequency
Nocturia
Dysuria (painful or difficult urination)
Urinary urgency
Incontinence
Urinary retention
Hematuria (blood in urine)
Pyuria (pus in the urine)
Oliguria (small amount of urine) or anuria (no urine)

Personal and family history

A history of past problems in the patient or a family member may aid in the establishment of a diagnosis.

High blood pressure

Renal dysfunction can affect any area of the body, as every organ system is dependent upon the kidneys' ability to filter the blood. Always look for symptoms related to other organ systems. High blood pressure may indicate an increase in body fluid volume related to the kidneys' inability to eliminate fluids.

Dialysis/kidney transplant

For family or personal history of acute or chronic renal failure leading to treatment, assess base problem surgical procedures to include date and outcome, frequency and type of dialysis, and current medications.

Urinary system cancer

Includes neoplasms of the kidney, renal pelvis, ureter, bladder, prostate, or urethra.

Cystitis/Pyelonephritis/urinary calculi

Cystitis may be acute or chronic inflammation of the bladder usually due to urinary tract infection (UTI). Kidney infection or pyelonephritis may also be acute or chronic. Symptoms include fever, chills, flank pain, frequent urination with burning.

Urinary calculi (stones) may occur anywhere in the urinary tract. The patient with calculi is in pain, and may have a urinary obstruction and infection.

Urination problems

A change in the pattern of urination may accompany disorders of the urinary system. Dysuria or difficulty in urinating includes difficulty starting stream, maintaining stream, burning or pain with urination. This may be due to infection or stricture of urinary tract. Burning on urination is usually associated with UTI or sexually-transmitted disease. Frequent urinating can be associated with diabetes, UTI, or renal failure. If the patient complains of urinating at night, or nocturia, obtain number of times patient gets up to urinate. Excessive nocturia indicates the kidney is unable to concentrate the urine.

Incontinence may be related to a congenital defect, UTI, decreased bladder capacity, urethral obstruction, damage to the central nervous system (stroke, spinal cord injury), use of some medications (diuretics, sedatives), or stress incontinence (common in elderly women) when urine escapes when coughing or laughing.

Urine

It is important to assess the characteristics of urine if a urinary system problem is present. The color of urine may be affected by medications, hydration status, or blood or pus in the urine. Urine with frank blood may be seen in UTI or renal calculi. Pus or cloudy urine may be seen with UTI.

Urinary system testing

Urinalysis or urine culture is commonly ordered on patients experiencing urinary problems, and it will be helpful to assess if the patient needs to void so a specimen may be obtained. Blood tests which may be ordered include blood urea nitrogen (BUN), creatinine, protein, albumin, potassium, sodium, calcium, phosphorus, bicarbonate, and glucose. Diagnostic testing may include cystoscopy, kidney-ureter-bladder x-ray (KUB), I.V. pyelogram (IVP), angiography, renal scan, cystourethrogram, cystometrogram, or biopsy.

Treatments and medications

When assessing medications, take side effects that could be contributing to the chief complaint. Many patients in renal failure may be fluid restricted while those with UTI may be encouraged to force fluids. Weight assessment is important as gain may be attributed to fluid retention.

URINARY SYSTEM ASSESSMENT FORM

Chief complaint

Patient's statement________________Onset______________Symptoms______________

Frequency__________Duration____________________Other areas affected____________

Have you had this before?________________________When?________________________

What have you done for this problem?____

What do you think caused this to happen?______________________________________

What changes have you had to make because of this problem?______________________

____

Personal and family history

	Patient	Family member		Patient Now	In past
High blood pressure	_____	_____	Cystitis	____	____
Dialysis/treatment	_____	_____	Pyelonephritis	____	____
Kidney transplant calculi	_____	_____	Urinary	____	_____
Urinary system cancer	____	____			

Urination problems: Dysuria____

Frequency_____________Urgency___________Incontinence_______Nocturia__________

Retention of urine________________Swelling of ankles/feet/hands__________________

Urine: color___________blood____________pus____________cloudy or clear_________

Urinary system testing

Urinalysis_____________Urine culture_____________24-hour urine collection_________

Blood testing____________________Cystoscopy_____________KUB_________________

IVP__________Renal angiography_______Renal scan__________Biopsy_____________

Cystourethrogram_____________Cystometrogram___________Other________________

Current treatment and medications

Name______________________________Dose________Frequency________Route_______

Name______________________________Dose________Frequency_______Route_______

Is your fluid intake restricted?________________How much fluid/day?_______________

What is your usual weight?_______________Have you lost/gained?________________

Are you on a special diet?_________What kind?____________________________________

Is there anything else you want me to know? _____________________________________

____

Physical Assessment

Assessing the urinary system is difficult due to the internal location of the components involved. The health history and laboratory results are an important part of urinary system assessment. The physical assessment of this system is done in the following order:

1. Inspection
2. Palpation
3. Auscultation

Inspection

General appearance/LOC

Does the patient appear ill, pale, flushed, or in pain?

Are they short of breath from abdominal distention?

Renal failure can lead to confusion, disorientation, and

coma, assess the level of consciousness

Skin

Skin that appears yellowish brown, dry, with multiple breaks or sores may be caused from the kidneys inability to remove waste products from the blood in end-stage renal disease. Pallor or flushed skin may be present. Record the location of any lesions or sores.

Edema

Edema usually starts in dependent areas and then may become generalized. Edema is not usually apparent until 5-10 pounds of excess fluid are present.

Describe location of edema such as ankles, feet, hands, periorbital, sacral, or anasarca (generalized edema).

Assess the severity; it is helpful to measure the girth of an edematous extremity to assess the effectiveness of future interventions. Edema can also be described on a scale of 1+ to 4+ edema 1+ is slight edema while 4+ is severe, pitting edema.

Halitosis/acetone breath

Acetone odor to breath may occur in end-stage renal disease. Some patients may complain of a persistent bad or metallic taste in the mouth.

Abdominal distention

Distention with a protuberant abdomen may occur in some kidney disease as fluid collects in the serous cavities. Bladder distention may occur with urinary retention.

Urinary catheter

If catheter is present list type and size, assess for patency and note if urine is draining. Assess that catheter is secured.

Urine characteristics

Urine may be described as colorless, straw-colored, yellow, amber, tea-colored, pale red, brown, or bloody. Urine should be clear, however, it may be cloudy if infection is present, or milky when alkaline and precipitated phosphates are present. Normal urine contains some sediment, however, with disease sediment is increased.

Palpation

Skin turgor/texture/temperature

Skin may be dry and cracked in renal failure. Over hydration and edema may be present in urinary retention

Striae may occur when skin stretches from ascites

Temperature may be elevated with UTI

Bladder/kidney

Kidneys are not normally palpable in healthy adults. If any pain is noted on palpation note the location and severity.

Auscultation:

Assessment of blood pressure should be done as elevations are common in kidney disease. Listen to the midline of the abdomen for bruits, if present, may indicate renal stenosis.

URINARY SYSTEM PHYSICAL ASSESSMENT FORM

Inspection:

General appearance________________________________LOC______________________

Skin________________________Halitosis/acetone breath____________________________

Edema: location______________________________severity________________________

Venous distention______________________Abdominal distention___________________

Body odor_____________Incontinence_____________Urinary dribbling_______________

Urinary catheter: type______________patency_______________draining____________

Urine: color____________clarity____________sediment__________malodorous________

Palpation:

Skin turgor_______________texture_______________temperature__________________

Bladder: distention___________Kidney: palpable/size__________ tenderness________

Pain over bladder_____________Flank pain____________Groin pain_______________

Auscultation:

Blood Pressure_____/_____Abdomen for bruit/murmur____________________________

Assessment Notes:

____

____

____

____

____

____

Section III – LABORATORY & DIAGNOSTIC TESTS

Table of Common Laboratory Tests

Test Name	Indications	Comments
Albumin (Serum) 3.5-5.0 g/dl or 52-68 of the total protein	Renal failure Nephrotic syndrome Edema Liver disease Ad-vance cancer Heart failure	**Regarding collection:** Part of protein electrophoresis Use red-top tube collect 5-10 ml **Results:** Decreased in renal failure, edema Assess urinary output if decreased Increased in dehydration
Blood urea nitrogen (BUN) 2.9-8.2 mmol/L Male 10-25 md/dl Female 8-20 mg/dl	Dehydration Renal failure Over-hydration	**Regarding collection:** Use gray-top tube collect 5 ml NPO for 6-8 hours preferable **Results:** Decreased in over-hydration Increased in dehydration and renal failure, and with some medications
Creatinine (Cr) 53-106 micro mol/L 0.6-1.2 mg/dl	Renal disease Urinary cancer Heart failure	**Regarding collection:** Use red-top tube collect 5-10 ml Some medications will interfere with test results hold these with physician permission for 24 hours prior to test-ing. List any drugs taken on laboratory slip. **Results:** Increased in renal disease Assess urinary output Usually com-pared with BUN
Electrolytes	Suspected imbalance	
Potassium (K+) Serum 3.5-5 mmol/L Urine 40-80 mmol over 24 hours	Renal disease Weakness Diuretic use IV Potassium use Trauma Surgery	**Regarding collection: (serum)** Use green or red-top tube, 5-10 ml Do not use tourniquet if possible hemolysis gives false-high result **Regarding collection: (urine)** 24 hour urine collection Obtain appropriate container Keep on ice or refrigerated **Results: (serum)** Decreased in renal tubular disease trauma, surgery, some diuretics In-creased oliguria, anuria, acute renal failure, IV potassium use

Sodium (Na+) 135-145 mEq/L 135-145 mmol/L	Dehydration Over-hydration Trauma Renal diseases SIADH	**Regarding collection:** Use green or red-top tube, 5-10 ml Usually collected with serum K+ **Results:** Decreased in SIADH, surgery, burns salt-wasting renal disease Increased in dehydration, high Na+ dietary intake
Phosphates (Hpo4, H_2PO4-) 1.7-2.6 mEq/L 2.5-4.5 mg/dl 0.78-1.52 mmol/L	Kidney dysfunction	**Regarding collection:** Use red-top tube collect 5 ml NPO for 4-8 hours except water **Results:** Increased in renal failure
Protein - Total 6.0-8.0 g/dl	Cancer Malnutrition Renal disease	**Regarding collection:** Usually done with albuim Use red-top tube collect 5-10 ml Prevent hemolysis of sample NPO 8 hours except water and drugs Avoid high-fat diet before test **Results:** Increased in dehydration, renal disease
Protein - Urine Random urine 0-5 mg/dl 24-hour urine 15-150 mg total	Renal disease Glomerular damage	**Regarding collection: (random)** Assess orders for type of specimen Random-midstream urine specimen Test with combistix Match results with bottle **Regarding collection: 24-hour** Obtain special container Discard first void Save all urine for 24 hours **Results:** Decreased diluted urine Increased chronic pyelonephritis, glomerulonephritis, renal disease
Urinalysis-routine	Renal diseases UTI System disease	**Regarding collection:** Collect fresh voided specimen Usually need 50 ml or more Assess orders for clean-catch, if ordered obtain container and wipes
Characteristic: Color straw or amber	**Results:** Colorless - diluted urine, diabetes insipidus, chronic renal disease, excess alcohol ingestion Orange - concentrated urine, fever, food coloring, increased biliburin Drugs: aminopyrine, furoxone, nitrofurantoin, phenazopyridine, sulfonamides	

	Red/brown - menstrual contamination, hematuria, beets, bilirubin Drugs: azogantrisin, dilantin, cascara, thorazine, doxidan, ex-lax Blue/green - Drugs: amitriptyline, methylene blue, methacarbamol Brown/black - lysol poisoning, melanin, bilirubin Drugs: cascara, chloroquine, iron injection
Appearance clear	Cloudy - infection, pus, WBC, RBC, phosphates, sperm Milky - fat, pyuria
Odor aromatic	Ammonia - Bacteria Foul - Bacteria (UTI) Sweet/fruity - Diabetic acidosis, starvation Mousey - Phenylketonuria
pH average 6 4.5-8 WNL	Decreased - acidosis, increased intake of cranberries, starvation, diarrhea, high protein diet Increased - UTI, Drugs: kanamycin, streptomycin sulfonamides, aspirin, diamox, HCO_3
Specific gravity 1.005-1.030	Decreased - inability to concentrate urine, diabetes, overhydration, renal disease, potassium loss Increased - dehydration, fever, IV albumin, diabetes
Protein	See protein
Glucose negative	Increased above 15 mg/dl - diabetes, stroke, infections, Cushing's, anesthesia, glucose IV Drugs: aspirin, ascorbic acid, keflin, streptomycin, epinephrine
Ketones negative	Increased 1+-3+ - Ketoacidosis, high protein-low carbohydrate diet, starvation
Red blood cells 1-2	Increased - menstrual contamination, trauma, renal or urinary disease Drugs: anticoagulants, aspirins
White blood cells 3-4	Increased - UTI, fever, renal disease

Table of Diagnostic Tests

Test Name	Indications	Comments:
Cystoscopy Normal urethra and bladder structure	Hematuria Chronic UTI Renal calculi Urinary tract tumor	**Pre-procedure:** Signed consent form is required A lighted cystoscope will be used to visualize bladder Pre-medication may be ordered Check orders for fluids before test Procedure takes about 1 hour **Post-procedure:** Assess VS, monitor for infection Monitor urine for output and clots
Intravenous urography (IVU) Intravenous pyelography (IVP) Cystometrogram to record bladder pressure may be done Normal structure, size, function of bladder, ureters and kidney	Urinary calculi Urinary tumors Polycystic kidneys Recurrent UTI	**Pre-procedure:** Tomography may be done with IVU Signed consent form is required NPO for 8-12 hours before test Laxative or enema may be ordered Pre-medication will be given Assess for allergies to iodine, seafood, or contrast dyes Record baseline vital signs Assess BUN, if greater than 40 test is usually not done, notify doctor **Post-procedure:** Monitor urine output and VS
Kidney, ureter and bladder x-ray (KUB) Normal size and structure of KBU	To determine size of kidneys, bladder Renal calculi Kidney or bladder mass	**Pre-procedure:** Procedure usually takes 10-15 min Assess for pregnancy and report **Post-procedure:** No activity restrictions
Renal angiography Normal renal blood vessel patency and structure	Renal tumor Renal cyst Renal aneurysms Assess renal artery Hypertension Renal failure	**Pre-procedure:** Signed consent is required Explain procedure-catheter will be inserted into femoral artery up to the renal artery for visualization NPO for 8-12 hours before test Pre-medication will be ordered IV, check orders Procedure takes 1-2 hours **Post-procedure:** Assess for bleeding, VS, output Bed-rest for 12-24 hours Assess site for hematoma Assess peripheral pulses and report

Section IV -- PROCEDURES AND CONDITIONS

Performing Urinary Catheterization

Urinary catheterization may be done for several reasons:

- To relieve urinary retention
- To obtain sterile urine specimen for testing
- To measure the amount of residual urine in the bladder
- To empty the bladder before surgical or diagnostic procedures

Equipment needed for catheterization:

1. Sterile catheter of appropriate size and type

Size	Type
14-16 French for female adult	Foley for continuous
18-20 French for male adult	Straight for single use
8-10 French for children	Drainage or sample

2. Sterile catheterization kit

   Gloves, antiseptic solution, cotton balls, forceps, lubricant, receiving container and sterile towel

3. Sterile collecting bag and syringe with sterile water if catheter is to remain in place

Steps:

Gather equipment and wash your hands

Observe Universal Precautions

Prepare the patient

- Provide for privacy and explain the procedure
- Position the patient and expose the genitalia
  - Female in lithotomy position
  - Male in supine

While maintaining sterility prepare catheterization equipment

- Open kit and don gloves
- Position drape
  - Females under buttocks
  - Males around penis
- For continuous catheter test balloon, instill and remove fluid
- Pour antiseptic over cotton balls
- Place lubricating jelly on catheter tip

Perform catheterization

- Using non-dominant hand expose the urinary meatus
  - Females separate labia majora and minora
  - Males hold penis with foreskin retracted

Holding cotton ball in forceps cleanse meatus

Females:
1 stroke with 1 cotton ball for each labia minora
Stroke from top to bottom once then discard cotton
1 stroke with 1 cotton ball to cleanse urinary meatus
Stroke from top to bottom then discard cotton
Males:
1 stroke with 1 cotton ball to encircle urinary meatus
Discard cotton and repeat

Insert catheter slowly, using dominant hand until urine flows

Females: Urethra is 4 cm or 1.5 inches (approximate)
Males: Urethra is 20 cm or 8 inches (approximate)

Stop insertion if obstruction is met, do not force catheter

Allow urine to drain into receptacle

For continuous bladder drainage

Release genitals and reattach syringe to balloon port
Fill balloon as recommended by manufacturer
Tug lightly on catheter to assure balloon is filled and holding catheter in place.

Connect to gravity collection bag with bag positioned below bladder level and tape tubing securely to thigh.

Common problems and solutions:

Difficulty locating urinary meatus in female patients

Become familiar with the location of the meatus BEFORE you try the first catheterization. A common error is insertion of the catheter into the vagina. The urethra is above the vagina. Some practitioners recommend placement of a cotton ball at the vaginal opening to prevent confusion.

Urine on bedding or patient

Always place the end of the catheter in a collecting basin prior to inserting the tip in the meatus.

Tubing detached from collection bag

Tape all connections. Tape bag to leg to prevent unnecessary tension on connections.

Discomfort during insertion

If discomfort is felt during insertion, have patient take slow steady breaths with deliberate exhalation, this promotes relaxation and lessens feelings of helplessness.

Section V -- DIETS

POTASSIUM RESTRICTED

Indications

Renal disease chronic renal failure

Comments

Potassium is restricted in renal disease when the kidneys are unable to excrete it normally. The excess potassium leads to hyperkalemia.

Amount of potassium allowed should be specified in physicians or diet orders.

Often used with sodium and protein restricted diets.

Restrictions

High potassium foods:

Vegetables	Fruits	Other
Artichokes	Avocados	Coffee
Asparagus	Dried apricots	Meats
Beans, lima	Bananas	Peanut butter
Beans, snap	Cantaloupe	
Broccoli	Oranges	
Brussel sprouts	Orange Juice	
Cauliflower	Dried prunes	
Corn	Raisins	
Mushrooms		
Potatoes		
Squash, winter		

Allowable Foods

Low potassium foods:

Vegetables	Fruits	Other
Beans, green	Cranberry juice	Breads
Carrots	Cranberry sauce	Spaghetti
Cucumbers	Blueberries	Cottage cheese
Lettuce	Lemons	Eggs
Radishes	Pears	Chicken
Romaine	Peaches	Lamb
Tomatoes	Tangerines	Fish
Pepper, green	Watermelon	Fats, butter

Other Interventions	Assess for hyperkalemia to include: Nausea, diarrhea, muscle spasms, muscle weakness, cardiac arrhythmia, cardiac arrest. Assess for hypokalemia to include: Tingling in extremities, tetany, muscle twitching, convulsions. Instruct patient to read label of any salt substitute used for amount of potassium.

PROTEIN RESTRICTED

Indications	Renal disease, chronic renal failure
Comments	The primary aim of protein restriction is to reduce the workload of the kidneys by reducing the amount of protein by-products Amount of protein restriction should be specified: Diet may be 20gm, 40gm, or 60gm protein diet Protein is needed by the body and cannot be eliminated totally Compliance is difficult to obtain Multi-vitamins are often prescribed to ensure adequate nutrition
Restrictions	Protein is restricted Meats, eggs, and milk products are limited 20 gm protein diet: All meats are restricted Allows for 1 egg white daily Allows for 3/4 cup of milk daily Protein rich breads, cereals and vegetables are restricted 40-60 gm protein diet: Some meats are allowed More milk is allowed Check with dietary department

Allowable Foods	Low protein foods:

Vegetables	Fruits	Fats
Green Beans	Apples	Butter
Cabbage	Blackberries	Margarine
Carrots	Blueberries	Sour cream
Cucumbers	Cherries	Oil
Egg plant	Grapefruit	
Lettuce	Peaches	
Onions	Pears	
Radishes	Pineapple	
Squash	Prunes	
Tomatoes	Strawberries	

Other items allowed: synthetic juices, hard candies, carbonated beverages, coffee and teas.

Other Interventions Low protein diets are often used with low salt or low potassium diets; use of foods should be checked for other dietary restrictions.

Section VI -- DRUGS

The tables below supply only general information, a drug handbook should be consulted prior to administering any unfamiliar drug.

DIURETICS

Action: To increase urine excretion

Indications: To promote fluid loss in renal disease, congestive heart disease, high blood pressure, liver malfunction, or other illness where edema is a symptom.

General comments: There are several types of diuretics. Some work to increase the glomerular filtration rate while others decrease the amount of fluid absorbed from the tubules allowing for more fluid to be excreted. Instruct patient to take medication exactly as ordered, and to increase potassium intake if diuretic is potassium wasting. Foods high in potassium can be found listed on page of the diet section of this chapter. Inform patient of symptoms of hypokalemia to include; fatigue. muscle weakness, confusion, cramps, and irregular pulse beat. Input and output of fluids should be monitored with use of these drugs. Blood pressure should be monitored and daily weights should be done on all patients. Diuretics are usually given in the morning. See also pages to for diuretic information.

Examples of drugs int this classification:

Generic	Trade	Comment
Acetazolamide	Diamox Acetolazam	Carbonic anhydrase inhibitor Given PO, IM, or IV Do not crush capsules Tablets may be crushed Avoid excessive salt intake Loop diuretic
Bumetanide	Bumex	Given PO, IM. IV Diuretic action 40X stronger than furosemide PO onset - 30-60 minutes IM onset - 40 minutes IV onset - less than 5 min Administer IV slow over 1-2 minutes to prevent ototoxicity

Ethacrynic Acid	Ederin Ederin Sodium	Loop diuretic Given PO, IV Mix with food to decrease stomach upset Administer IV slow over 5 minutes to prevent ototoxicity
Furosemide	Lasix Furoside	Loop diuretic Given PO, IV, IM Novosemide PO Onset in 30-60 minutes IV onset in 5 minutes Administer IV slow no more than 4 mg/minute
Indapamide	Lozol	Athiazie-like diuretic Given PO Monitor BUN, creatitine, UNC acid
Mannitol	Osmitrol	Osmotic diuretic Given IV Dose dependent upon output Desired output 3-50 ml/hr Assess output at least hourly
Spironolactone	Aldactone	Potassium-sparing diuretic Given PO Administer with food for best absorption

SECTION VII – GLOSSARY

Acetone breath	Breath with a sweet, fruity odor usually due to ketoacidosis
Anasarca	Severe, generalized edema
Anuria	No urine output
Azotemia	Urea and other nitrogenous wastes in the blood; occurs in renal failure
Calculi	Stones, plural form of calculus, usually composed of mineral salts; may occur in kidney, bladder or ureters. May block urinary flow leading to pain and inflammation
Calculus	Single stone
Cystitis	Inflammation of the bladder; symptoms include dysuria and frequent voiding
Cystocele	Herniation of the bladder into the vagina; may occur to injury during childbirth
Dialysis	The process of cleansing blood in patients with severe renal disease. Blood is passed through a tube that serves as a membrane where waste products are removed
Dysuria	Difficult or painful urination
Edema	Swelling in the body
Enuresis	Incontinence of urine
Epispadias	Congenital abnormality of the urinary meatus. Meatus is not found in usual location. May occur in males or females
Filtrate	Fluid that has been filtered from the glomerular blood into the kidney tubule (another possibility would be into Bowman's capsule)

Halitosis	Bad breath
Hematuria	Blood in urine
Hypospadias	Male congenital abnormality of the urinary meatus. Meatus is located on the underside of the penis
Ketones	An acidic end product of fat metabolism
Micturition	The act of urinating
Nephrolith	A kidney stone or calculus
Nephrectomy	Surgical removal of a kidney
Nephron	The functional unit of the kidney; consists of a glomerulus and tubules. Used to filter wastes from blood and selectively reabsorb substances
Nocturia	Urinating during the night
Oliguria	Scanty urine output less than 30 ml/hr
Polyuria	Excessive urination
Pyuria	Pus in the urine
Stomatitis	Inflammation of the mouth
Uremia	Azotemia, urea, and other nitrogenous wastes in the blood; usually occurs in renal disease
Uresis	Normal urine output, normal urination
Urinary tract infection (UTI)	Urinary tract infection

APPENDIX 1
THE BASIC ASSESSMENT

General information:

- Name, age, sex, marital status, race
- Referring physician
- Source of information (patient, parent, significant other)
- Persons present
- Known allergies
- Name of translator, if needed, and native language

Example:
Jeremy Smith, 16-year-old black male, single
Referred by Dr. C. O'Niell
Information obtained from parents: Bob and Neta Smith
Persons present: Both parents and patient

Chief complaint:

- Stated by the patient in a direct quote
- Concise and brief statement
- Recorded in quotation marks if direct quote
- Time reference

Example
Parents state "blacked out after breakfast and started an epileptic fit." Lasted until EMS arrived.
Patient states, "Dr. O'Niell sent me over for a CAT scan."

Present illness:

- What happened?
- When did it begin?
- How did it begin?
- Has the problem changed over time? Why does the patient think it changed?
- What treatments have been given?
- Medications: dose, frequency, result?
- What physicians has the person seen and with what result?
- What are the symptoms right now?

- Chart in chronological order whenever possible

  Example:
  States seizure activity began after breakfast, and involved only the right side of the body. This is the third seizure in seven years, two previous seizures (May 1980 and November 1986) involved the entire body and were related to high temperatures. No anticonvulsants were prescribed. A CT scan in November of 1986 was normal, ordered by Dr. R. Graham of Atlanta General Hospital. At present the only complaint is headache and generalized weakness on the right side.

Past medical history:
- Illnesses, hospitalizations (include name), surgeries
- Current medications
- Allergies
- Past blood transfusions and reactions, if any
- Tobacco, alcohol and drug use
- Significant childhood illness

Social history:
- Educational level
- Marital status and/or history
- Vocation, work hours, occupation hazards
- Usual sleep pattern, routines, elimination patterns

Family history:
- Members of immediate family to include age, sex and general health status
- Cause of death for any deceased, immediate family members
- History of cancer, heart disease, lung, kidney or neurological disorders in blood relatives
- Determine if other family members have similar problems

Review of symptoms:
- Each of the major body systems are reviewed for past problems
- See assessment forms for each system
- Includes a general section where weight changes, fever, chills and allergies can be documented

Interview tips:

- Use open ended questions to obtain information. Example: "Tell me about your pain."
- Avoid judgmental questions such as, "You don't have more than two drinks a day do you?"
- Ask visitors to step out of room to provide confidentiality and to avoid embarrassing the patient
- If possible, provide quiet environment for the interview with few distractions, turn off television, close the curtain

Physical examination tips:

The physical examination is composed of several factors that can be equally important. These are observation, palpation, and auscultation.

- Always provide for privacy prior to the exam. Ask visitors to step outside, pull the curtain or shut the door
- Always leave the patient properly gowned, assisting in tying gowned and putting on pajamas
- Begin with the presenting problem, unless a full head-to-toe physical is to be performed

Example:

- If your patient has just returned from surgery, check the dressings, bowel sounds, level of consciousness, respirations, blood pressure, and lung sounds first. If surgery was for a brain tumor you would concentrate on the neurological area for reduction of a fracture you would assess peripheral pulses of affected extremity
- Chart areas assessed that are of significance; negative assessments can be just as vital as positive ones

Example:

A negative Babinski is an important finding in a child with a suspected brain tumor and should be charted if performed.

BIBLIOGRAPHY

Books

Bates, Barbara: **A Guide to Physical Examination,** 5th. ed., Philadelphia, J.B. Lippincott Co., 1991.

Benenson, Abram S.: **Control of Communicable Diseases in Man,** 15th ed., American Public Health Association, 1990.

Berkow, Robert, editor-in chief: **The Merck Manual,** 15th ed., Rahway, N.J., Merck Sharp & Dohme Research Laboratories, 1992.

Carpenito, Lynda Juall: **Handbook of Nursing Diagnosis,** 4th edition, Philadelphia, J.B. Lippincott Company, 1991.

Cloherty, John P. and Ann R. Stark: **Manual of Neonatal Care,** Boston, Little, Brown and Company, 1991.

Gomella, Leonard G., G. Richard Braen and Michael Olding: **Clinician's Pocket Reference,** 5th ed., Norwalk, Connecticut, Appleton-Lange, 1991.

Govoni, Laura E. and Janice E. Hayes: **Drugs and Nursing Implications,** 6th ed., Norwalk, Connecticut, Appleton-Lange, 1992.

Kee, Joyce LeFever: **Laboratory and Diagnostic Tests with Nursing Implications,** 2nd ed., Norwalk, Connecticut, Appleton-Lange, 1992.

Lebovitz, Harold E., editor: **Therapy for Diabetes Mellitus and Related Disorders,** Alexandria, Virginia, American Diabetes Association, Inc., 1991.

Malasanos, Lois et al: **Health Assessment,** St. Louis, Mosby-Yearbook Company, 1990.

Metheny, Norma M.: **Fluid and Electrolyte Balance - Nursing Considerations,** Philadelphia, J.B. Lippincott Company, 1992.

Phipps, Wilma J., Barbara C. Long and Nancy Fugate Woods: **Medical-Surgical Nursing, concepts and clinical practice,** St. Louis, Mosby-Yearbook Company, 1991.

Poole Arcangelo, Virginia: **Weaver & Koehler's Programmed Mathematics of Drugs and Solutions,** 5th ed., Philadelphia, J.B. Lippincott Company, 1992.

Porth, Carol: Pathophysiology, **Concepts of Altered Health States,** 3rd ed., Philadelphia, J.B. Lippincott Co., 1990.

Schroeder, Steven A, Marcus A. Krupp and Lawrence M. Tierney Jr.: **Current Medical Diagnosis & Treatment,** Norwalk, Connecticut, Appleton-Lange, 1992.

Taber's Cyclopedic Medical Dictionary 17th ed., edited by Clayton L. Thomas, Philadelphia, F. A. Davis Company, 1991.

Timby, Barbara K., **Clinical Nursing Procedures**, Philadelphia, J.B. Lippincott Company, 1989.

Weldy, Norma Jean: **Body Fluids and Electrolytes, A Programmed St. Louis Presentation,** 6th ed., St. Louis, Mosby-Yearbook, 1992.

Whaley, Lucille F. and Donna L. Wong: **Nursing Care of Infants and Children,** 2nd ed., St. Louis, Mosby-Yearbook Company, 1991.

Journals

Anastasi, Joyce K. and Julie Linksman Rivera: "AIDS drug update, ddI and ddC", **RN,** November, 1991.

Barrick, Bill: "Light at the end of a decade", **AJN**, November, 1990, pages 37-40.

Bavin, Terry K. and Marjorie A. Self: "Weaning from intra-aortic balloon pump support", **AJN,** October, 1991.

Bockus, Sherry: "Troubleshooting your tube feedings", **AJN,** May, 1991, pages 24-28.

Braun, Anne E.: "Drugs that dissolve clots", **RN**, June, 1991, pages 52-57.

Carroll, Patricia: "What's new in chest-tube management", **RN,** May, 1991, pages 34-40.

Clark Mims, Barbara, "Interpreting ABGs", **RN,** March, 1991, pages 42-46.

Fitzgerald, Margaret Ann: "The physical exam", **RN,** November, 1991, pages 34-38.

Gehring, Patsy Eileen: "Physical assessment begins with a history", **RN,** November, 1991, pages 27-31.

Greifzu, Sherry: "Helping cancer patients fight infection", **RN**, July, 1991, pages 24-28.

Hefti, Deanne "Chest trauma", **RN**, May, 1991, pages 28-32.

Hill, Martha N. and Carlene Minks Grim, "How to take a precise blood pressure", **AJN,** February, 1991, pages 38-42.

Hoffman, Leslie A., Marion C. Mazzocco and James E. Roth "Fine tuning your chest PT" **AJN**, December, 1987, Vol 87 No 12, pages 1566-1573.

Howard, Patricia: "Elevated cholesterol: A nurse's guide to drug therapy", **RN**, August, 1991, pages 26-29.

Meyer, Charles: "Nurses vote for the most valuable new drugs", **AJN,** September, 1991, pages 33-37.

Meyer, Charles: "Nursing and AIDS: A decade of caring", **AJN**, December, 1991, pages 26-31.

National Institutes of Health: "The 1988 Report of the Joint, National Committee on Detection, Evaluation, and Treatment of High Blood Pressure", **NIH Publication** No. 88-1088, US Government Printing Office.

Solomon, Jacqueline: "Managing a failing heart", **RN**, August, 1991, pages 46-50.

Stiesmeyer, Johanna K.: "What triggers a ventilator alarm?", **AJN**, October, 1991.

"AIDS: A guide for survival," The Harris County Medical Society and The Houston Academy of Medicine, 1987.

Yacone-Morton, Linda Ann: "Cardiac assessment", **RN**, December, 1991, pages 28-34.

INDEX